# THE TVA REGIONAL PLA
# DEVELOPMENT PRC

# The TVA Regional Planning and Development Program

## The Transformation of an Institution and Its Mission

AELRED J. GRAY
*Former Chief of Regional Planning, Tennessee Valley Authority*

*and*

DAVID A. JOHNSON
*Professor Emeritus, The University of Tennessee*

Routledge
Taylor & Francis Group

LONDON AND NEW YORK

First published 2005 by Ashgate Publishing

Published 2017 by Routledge
2 Park Square, Milton Park, Abingdon, Oxfordshire OX14 4RN
711 Third Avenue, New York, NY 10017, USA

First issued in paperback 2017

*Routledge is an imprint of the Taylor & Francis Group, an informa business*

**British Library Cataloguing in Publication Data**
Gray, Aelred J.
  The TVA regional planning and development program : the
  transformation of an institution and its mission. - (Urban
  planning and environment)
  1.Tennessee Valley Authority - History 2.Regional planning
  - Tennessee River Valley - History
  I.Title II.Johnson, David A., 1935-
  307.1'09768

**Library of Congress Cataloging-in-Publication Data**
Gray, Aelred J., (Aelred Joseph) 1909-2000
  The TVA regional planning and development program : the transformation of an
  institution and its mission / Aelred J. Gray and David Johnson.
    p. cm. -- (Urban planning and environment)
  Includes bibliographical references and index.
  ISBN 0-7546-3786-7
    1. Tennessee Valley Authority--History. 2. Regional planning--Tennessee River Valley.
  3. Organizational change. 4. Corporate culture. I. Johnson, David A., 1935- II. Title. III.
  Series.

  HD9685.U5G74 2004
  307.1'09768--dc22
                                                                    2004046268
  ISBN 13: 978-1-138-25880-8 (pbk)
  ISBN 13: 978-0-7546-3786-8 (hbk)

# Contents

# List of Figures

# List of Tables

# Preface

The Tennessee Valley Authority was created by Act of Congress in 1933. TVA was intended to be a centerpiece of President Franklin D. Roosevelt's New Deal response to the economic distress of the Great Depression. The Authority was conceived as a way to pull one of the poorest parts of the country out of poverty and to serve as a model for the planning and development of other regions of the country. It was designed to have the material support and legal authority of the federal government while operating with the flexibility of a private corporation. TVA was given three discrete missions, each focussed on the Tennessee River which flowed through seven states of the Old Confederacy. These were flood control, improved navigation, and most important, hydroelectric power generation. These missions were conceived as part of a larger, and more vaguely defined, objective of integrated regional development – social, economic and physical. The history of TVA is in essence the interaction and conflict between its three specific missions and its larger mission – lifting its region (defined in various ways) out of economic backwardness into a prosperity that eluded all of the country in the Depression. The challenge was to do so through intentional integrated regional and community planning. It was an enormous challenge.

Because of the vagueness of its charge to carry out regional development TVA never was able to provide a comprehensive regional plan for its river basin. Nevertheless, TVA had a profound impact on the Tennessee River Valley and on the American South as a whole. Perhaps as important, TVA helped win the World War which ended the Depression. The electricity TVA generated by harnessing the river through an extraordinary system of dams provided the energy for the aluminum production needed to manufacture military aircraft. The secret Oak Ridge atomic facility could only have been constructed where great power surpluses were available. TVA's hydro power refined the uranium that ultimately brought the War in the Pacific to a climax with the bombing of Hiroshima and Nagasaki.

There have been numerous histories of TVA, some synoptic, others focused on specific topics. Some were written by such important figures as Arthur Morgan and David Lilienthal, chairmen who guided the TVA in its early days. Other books and articles have focussed on political, economic and environmental impacts. This book supplements those works. Its claim to originality is primarily that it is the memoir of an extraordinary, dynamic staff member who served TVA in important positions for more than 30 years, nearly from its creation to late into the 20th century. Few people insideTVA

**Frontispiece:   Aelred J. Gray, Chief Regional Planner, Tennessee Valley Authority, at the beginning of his career**

*Source:* TVA files.

enjoyed the vantage point and the longevity of Aelred J. Gray. A.J. or 'Flash', as he was known from his Notre Dame football days, was in a unique position to observe and participate in the workings of TVA from its inception until his retirement in 1973. His professional career thus provides a unique opportunity to understand the workings of TVA over 40 years, providing a lens through which to view what TVA was supposed to be and what it became as an organization – its achievements, and its shortcomings. At the same time we need to recognize that Gray was looking at the day-to-day operations of TVA from a particular vantage point in the agency – that of an urban and regional planner. Planning and development was what TVA was supposed to be about. But as David Lilienthal liked to point out, TVA had no 'regional master plan'. What there was, was a general understanding of directions, a plan always in development, a visioning process, loosely defined, but one which was supposedly generally understood and one which was based on grassroots citizen and local and state government participation. As subsequent analysts demonstrated, TVA was in reality a federal agency imposing its will on the region, at least in its three primary missions – flood control, navigation, and electric generation. It could not have been otherwise. But it was in the area of community and local development that TVA had some discretion to respond to local needs, problems and opportunities, within a region-wide framework. It is this part of TVA's story that this work focuses on, through the work and views of the planner who directed this part of TVA's program for nearly three decades.

The community planning development story wihin TVA cannot be told without relating it to the larger context of the agency as a whole. To that extent the current work is a history of TVA from 1933 to 1994. But it is admittedly and unabashedly told from a particular point of view. Flash Gray, as those who knew him can attest, was a man of strong opinions. These opinions show up in this book. Flash was disappointed in the directions that the TVA planning program took in his final years. TVA, like other public agencies, changed significantly as conditions and organizational leadership changed. Flash felt that the agency had lost sight of its regional outlook, and had forgotten its institutional history. Critics and some of the TVA participants whose actions are discussed here may take the view that the tranfsormations that TVA went through were inevitable given the political realities, nationally, and in the Valley. I leave it to the reader to decide whether Gray's argument is persuasive. But persuasive or not, they are the lively recollections of a man who was on the scene. These memoirs were recorded over a period of several years during Mr. Gray's retirement and were based on his extensive personal files and the TVA archives. Flash Gray died on July 16, 2000. During his lifetime he had received numerous honors. In 1989 he won the Lifetime Achievement Award from the Tennessee Chapter of the American Planning Association. A few years later he was cited as a 'planning pioneer' by the American Institute of Certified Planners.

As editor of his writings I have revised as conservatively as I felt I could, preferring to let Mr. Gray speak for himself. The introductory and concluding chapters are my own contribution, but were reviewed in early draft form by Mr. Gray. I take responsibility for the content of the entire text with the caveat that I do not necessarily agree with all the judgements made by 'Flash' Gray. But I do agree with most of them.

David A. Johnson

# Acknowledgements

I would like to express appreciation for the assistance and support provided by a number of people and agencies in the preparation of this work. To Mrs. A.J. Gray, special thanks. Rose was helpful to the authors on many occasions, as we went through countless files and reports. H. Brown Wright read the manuscript and made many valuable suggestions. The TVA Communications Staff and the TVA Historian, Patricia Bernard Ezzell, were of great assistance in locating materials in the TVA files. Daniel Schaffer, former TVA historian, stimulated my interest in the early days of the agency. Donald Miller reveiewed the text and made numerous recommendations for improvement. He also shared with us the memoir prepared by his father, Harold Miller, an outstanding planner who served TVA with distinction for many years. Stephen Smith generously provided a valuable perspective on the recent history of the agency. It should be emphasized that the responsibility for what is in our narrative remains solely with the authors. At Ashgate Publishing, I was ably assisted by Valerie Rose, Carolyn Court and Rosalind Ebdon. Finally, I would like to thank my wife, Eleanor Stephens Johnson, who knew and worked with A.J. Gray many years before I ever did. Eleanor's insights and patience were, as always, indispensable.

David A. Johnson
Asheville, North Carolina

Chapter 1

# Introduction:
# The Emergence of Regional Planning
# Agencies in the United States

Regions have long been used to describe and understand significant facts and differences about the world, its climate, physical characteristics, natural resources, and demographic, social, and economic conditions. The regional geographers were at the forefront of this type of research. The United States with its varied climatic, resource, and physical conditions has been a fertile field for regional research: Baker's studies of agricultural regions; Fenneman's work on physiographic regions; and Marburg's soil classifications, are but a few of the many examples of early regional geographic research in the United States.[1] There also has been a long interest in creating regions for effective administration of developmental programs to deal with problems which transcend or cross boundaries of traditional governmental units such as states, counties, or cities. However, only in the twentieth century have efforts been made in the United States to organize planning and development agencies primarily on the basis of defined regional characteristics and problems.

Although regions have been an important vehicle for research into area problems, the place of regions in the traditional governmental structure and particularly the place of regions used for planning and development has been less clear. Because regions do not have a clear and recognized place in the governmental structure, regional institutions have difficulty in relating to the traditional governmental institutions with whom they share power. As a result the institutional relationships of regional agencies have been established by a process of trial and error. The Advisory Commission on Intergovernmental Relations (ACIR) in its 1972 report *Multistate Regionalism* took the position that because of the brief experience in regional planning and development operations a broad assessment of the effectiveness of such agencies could not be made. However, the Tennessee Valley Authority (TVA) has now been in existence for over seventy years and there exists a record of what TVA did over time as a regional development agency. Thus the occasion seems right for such an assessment. The same is true for other regional agencies and perhaps the time is right for studies of such institutions so that their effectiveness may be assessed and some principles and guidelines adopted for establishing new regional development agencies or for changing existing ones to meet new problems.

This study deals with TVA's activities as a regional development agency over the past seventy years. It is an examination of the TVA programs which were concerned with regional development, how they were related, the type of organization used, and the professional staffs which largely determined the scope of the regional program. The changes that took place over time are compared in three twenty-year periods beginning in 1933. The material used in the study is generally based on TVA materials such as internal memoranda, annual reports, publications, and the personal experiences of those involved in the regional development program. Other than noting the broad demographic and economic changes which have occurred in the Tennessee Valley Region and the Southeast the study does not attempt to relate TVA activities to these changes; such relationships have been well covered by many other studies. This study does however raise questions relating to TVA's response to these regional changes.

## Note

1    For a historical review of regionalism in the US see National Resources Committee, *Regional Factors in National Planning*, Dec. 1935; Advisory Commission on *Intergovernmental Relations, Multistate Regionalism*, April, 1972, and Advisory Commission on Intergovernmental Relations, *Regional Governance, Promise and Performance*, May 1973. See also: Friedmann, J, and Weaver, Cm (1979)

Chapter 2

# The First Twenty Years: The Formulation of the TVA Approach to Regional Planning and Development, 1933–1953

## Introduction

Although many regional agencies have been created over the past 50 to 60 years, only a few have been successful and have survived over an extended period of time.[1] In part this lack of success results from the special problems that regional agencies encounter both in the preparation of plans and in their implementation. The most important of these problems are: the definition of issues of regional concern as distinct from what are primarily concerns of other governmental units; the delineation of a physical region or regions to address regional concerns; the relation of the regional agency to traditional governmental units in matters of program planning and implementation; and the ability of the regional agency to adapt successfully over time, as regional needs and problems change. The Tennessee Valley Authority (TVA), generally regarded as the most important regional planning and development experiment in U.S. history, has been in existence for some seventy years and during this period has had to face and deal with all of these problems.

TVA's history falls into three distinct periods: the beginnings and first twenty years of operations – from the 1930s to the mid-1950s; consolidation of operations and completion of the hydro system – from the mid-1950s to 1975; and the period of maturity – 1975 to the present. This section deals with the first period – the historic background and origins of TVA and the initial twenty years of its operations – from the 1930s to the early 1950s. The section looks at how TVA defined areas of regional concern, the regions used in its planning and development programs, and how the agency related to other federal agencies and the state and local governments within the larger Tennessee Valley region.

## The Origins of the TVA Idea

TVA was one of Franklin D. Roosevelt's responses to the Great Depression of the early 1930s. But TVA's roots go back even further to the social and economic climate of the 1900s – populist, anti-robber baron, anti-private power monopoly, and anti-big city. Out of the political ferment of those times came the conservation movement, new approaches to rural development, and urban and regional planning. Although there had been efforts to apply the ideas of the conservation movement for integrated resource development, TVA was the first of the river basin or other large regional agencies to attempt to bring together the then new concepts of integrated water-resource development which was central to the conservation ideas of the time and regional planning which was a part of the new city and area planning movement.

For over one hundred years the people of the Tennessee Valley had been trying to improve the economy and reduce the isolation of the region by Tennessee River improvements which would permit dependable navigation from Knoxville to the Ohio river. Beginning in 1824 there was a series of attempts by both the Federal Government and the state of Alabama to overcome major impediments to navigation, particularly at Colbert Shoals and Muscle Shoals in north Alabama, and the Hales Bar area below Chattanooga. All these efforts produced few if any permanent improvements.

Two actions by the federal government were to lead to the first real improvements of the Tennessee River and have a major effect on both the creation and early operations of TVA. The first was the construction of Wilson Dam, built during World War I to supply power to two federally-owned nitrate plants. The second was Congressional authorization during the 1920s for the US Army Corps of Engineers to conduct comprehensive surveys of the water resources of the Tennessee River Basin. The reports of the Corps of Engineers compiled data on the water resources of the Tennessee River Basin and outlined alternative plans: high versus low systems of dams for the improvement of the river for navigation, flood control, and power generation (Schaffer, 1984).

### The TVA Act – The Legislative Basis for the Regional Development Program

Shortly after the end of the World War I, a major national debate developed over the use of Wilson Dam and the nitrate plants located near Muscle Shoals, Alabama either by a public entity or by private industry. Senator George Norris of Nebraska, a Progressive and long a proponent of public power, led the fight to oppose efforts to turn these valuable public facilities over to private industry. Norris proposed that these facilities be part of a plan for the development of the entire Tennessee River which would be based on the principles of integrated resource development. The surveys of the Corps of Engineers provided a base for Senator Norris's proposal and, as will be noted later, were a major factor in TVA's ability to move rapidly to prepare a plan for the improvement of the Tennessee River. Senator Norris introduced legislation in the 1920s for multi-purpose public development of the Tennessee River which was passed by the Congress but vetoed by both presidents Coolidge and Hoover.

The New Deal gave Senator Norris another opportunity to promote public power as a part of integrated resource development for the Tennessee River Basin. The newly-elected President, Franklin D. Roosevelt, was a natural ally. Through conversations with his uncle, Frederic Delano, and Charles D. Norton, who had fostered both the 1905 Plan of Chicago and the 1929 Regional Plan of New York and Its Environs, FDR became intrigued by the concept of regional planning as a framework for dealing with many of the nation's urban and rural problems (Johnson, 1996). While Governor of New York State (1929–1932), he learned of some of the new ideas about regional planning germinated in the discussions among members of the Regional Planning Association of America (RPAA). He was also familiar with the New York State Plan, which two RPAA members, Clarence Stein and Henry Wright, had prepared for his predecessor, Governor Al Smith. These associations expanded his thinking beyond physical planning of a single city to regional planning, which he saw as a tool by which the state could adopt policies relating to land use and population distribution (Roosevelt 1936; 116, 493, 496).

FDR's early speeches and papers clearly reveal a deep-seated and long-standing concern with farm problems, rural poverty, and the inadequate methods of providing services to rural areas. He also saw these problems as coupled with those of overcrowding and poor living conditions for the working people in the big cities, which he believed were fueled in part by the migration from farms to cities. The emerging concepts of regional planning seemed to FDR to provide the linkage between the two problems. FDR made clear his support for regional planning in an enthusiastic speech to an historic week-long conference on regionalism held at the University of Virginia in July 1931 (Roosevelt 1931).

It is within this context that the regional planning ideas of FDR were appended to the ideas of Senator Norris for a multi-purpose plan for the Tennessee River, which had focused on flood control, navigation, and power generation. The main body of the TVA Act as introduced by Senator Norris on 9 March 1933 consisted of Norris's comprehensive river improvement program, which had been vetoed earlier by President Hoover, plus the new Sections 22 and 23 which provided for surveys of regional development problems and for recommendations to Congress on needed legislation to help solve these problems.

Shortly after this bill was introduced in the Senate, two city planners, John Nolen and Frederick Gutheim, met with Senator Norris and proposed changes in the language of Sections 22 and 23, which were to become the regional planning sections of the TVA Act. These additions were accepted by Senator Norris and certainly had the encouragement and blessing of President Roosevelt (Augur, 1942, 1943).[2]

The river improvement sections of the TVA Act were very explicit and provided for a navigation channel from Knoxville to the Ohio River, flood control to protect valley cities and to help control floods on the Mississippi River, and power generation. In contrast, Sections 22 and 23 were very general and thus susceptible to either broad or narrow interpretation. If Sections 22 and 23 had been explicit as to the nature of the regional plan, as was characteristic of the planning legislation of the time, many of the conflicts and disagreements about the TVA's regional development program might have been avoided. Certainly, these sections did not outline how TVA was to implement regional plans or what power TVA had to implement such plans. It is also important to note that Sections 22 and 23 did not stand alone but were related to the other specific duties assigned to TVA. Sections 22 and 23 of the TVA Act follow:

Sec. 22. To aid further the proper use, conservation and development of the natural resources of the Tennessee drainage basin and of such adjoining territory as may be related to or materially affected by development consequent to this Act, and to provide for the general welfare of the citizens of said areas, the President is hereby authorized, by such means or methods as he may deem proper within the limits of appropriations made therefor by Congress, to make such surveys of and general plans for said Tennessee basin and adjoining territory as may be useful to the Congress and to the several states in guiding and controlling the extent, sequence, and nature of development that may be equitably and economically advanced through the expenditure of public funds, or through the guidance or control of public authority, all for the general purpose of fostering an orderly and proper physical, economic, and social development of said areas; and the President is further authorized in making said surveys and plans to cooperate with the States affected thereby, or subdivisions or agencies of such State, or with cooperative or other organizations, and to make such studies, experiments, or demonstrations as may be suitable to that end (48 Stat 69, 16 U.S.C. SEC.831u.).

Sec. 23. The President shall, from time to time, as work provided for in the preceding section progresses, recommend to Congress such legislation as he deems proper to carry out the general purposes stated in said section, and for the especial purpose of bringing about in said Tennessee drainage basin and adjoining territory in conformity with said general purposes (1) the maximum amount of flood control; (2) the maximum development of said Tennessee River for navigation purposes; (3) maximum generation of electric power consistent with flood control and navigation; (4) the proper use of marginal lands; (5) the proper method of reforestation of all lands in said drainage basin suitable for reforestation; and (6) the economic and social well-being of the people living in said river basin (48 Stat 69, 16 U.S.C. SEC . 831v.).

The power to carry out these two sections was given to the President and on 8 June 1933 FDR issued Executive Order 6161, which provided the TVA Board with the authority to use these sections of the TVA Act to prepare a regional development program:

In accordance with the provisions of Sections 22 and 23 of the Tennessee Valley Authority Act of 1933, the President hereby authorizes and directs the Board of Directors of the Tennessee Valley Authority to make such surveys, general plans, experiments, and demonstrations as may be necessary and suitable to the proper use, conservation, and development of the natural resources of the Tennessee River drainage basin and of such adjoining territory as may be materially related to or materially affected by the development consequent to this act, and to promote the general welfare of the citizens of said area; within the limits of appropriations made therefor by Congress (Roosevelt, 1936).

The generality of Sections 22 and 23 and of the executive order would seem to give the TVA Board of Directors great latitude in deciding how these sections were to be applied to a development plan for the Tennessee Valley. But this assumed that each member of the TVA Board had the same understanding as to what these sections intended TVA to do in preparing development plans and programs for the region (Gray and Johnson 1987).

Tracy Augur, the senior staff member of the Division of Land Planning and Housing, had researched the origins of Sections 22 and 23. (see Appendix for the memoranda describing the results of Augur's research). Writing in 1946 he expressed the view that these sections 'do not confer authority to do anything about development of resources except to make surveys and plans and incidental experiments and demonstrations. The powers granted by these sections of the Act are advisory powers and not developmental ones. The sections are planning sections and were so intended by the authors' (Augur, 1946). He goes on to state that in his opinion TVA powers for resource development are more accurately stated in the three opening paragraphs of the 1940 unpublished TVA report 'Regional Development in the Tennessee Valley'. These three paragraphs follow:

The statute creating the Tennessee Valley Authority is response to several interrelated problems traditionally national in character and interest. Far from being a novel excursion of the Congress of the United States, the Tennessee Valley Authority Act results from a legislative background of unusual maturity and is deeply rooted in a century of American history. The uniqueness of the statute is not in the problems toward which it was addressed, but in the method devised to meet these problems – that of a public corporation charged at once with the conduct of certain specific activities of the national government on a regional scale and with the responsibility of using this experience to assist in further regional development.

The positive acts which the Authority is authorized to perform are specific limited, and of long exceptance. It is directed to construct a series of dams on the Tennessee River and its tributaries to make the 650 miles of the main river navigable and, at the same time, effective in flood control; to distribute and sell the water power created by these dams in amounts consistent with their operation for navigation and flood control; to utilize the Muscle Shoals properties for the experimental manufacture and distribution of fertilizers; and to maintain these properties in stand-by condition for national defense purposes. In addition to this specific program of action, the Authority is directed to make surveys, demonstrations, and recommendations for legislation in order to promote further the general development of the region and to cooperate towards this end with existing public agencies – local, state, and federal. In these latter aspects, it should be noted that the Authority is only an advisory and cooperating body. It can survey, demonstrate, and recommend to other interests properly authorized to act; it cannot, however, without congressional authorization execute its own recommendations.

In essence the Authority's specified operational functions constitute an integrated approach to the interrelated water-resource problems of an entire watershed. By and from a considered application of this integrated approach to water control in the Tennessee Valley, it has been possible to derive benefits even above those originally contemplated and to discover opportunities for further benefits which have been translated into the life of the region through cooperation with local, state, and federal agencies. In this manner, regional development beyond the utilization of the water resource is being fostered in the Tennessee Valley. It is a continuing process in which local institutions and state agencies cooperate with the federal government on problems of mutual concern, This unity of interest and effort leads, to regional accomplishments not otherwise attainable (TVA, 1940).

Tracy Augur's conclusions concerning these sections would seem to be correct – namely that TVA did not have any significant regional development powers. This conclusion also seems to be supported by Erwin C. Hargrove who observed in his introduction to *TVA: Fifty Years of Grass-Roots Bureaucracy* that TVA 'is suspended at a mid-point between the national government in Washington and state and local governments' (Hargrove and Conkin, 1983). This important fact is missed or overlooked by most friends and foes of TVA. As a regional agency it is necessary to define with some degree of specificity the problems or opportunities unique to the region also as defined. This becomes even more important in a government such as we have in the United States where powers are shared between the different levels of government. All these issues were faced by the original TVA Board as it debated and attempted to define the approach to regional development and the content of the regional program.

### The TVA Board's Quest for a Regional Plan and Development Program – Differing Views and Interpretations of the TVA Act

The TVA Act placed control of the Tennessee Valley Authority in a board composed of three directors selected by the President and confirmed by the Senate. One of the directors was to be designated chairman of the board, but his voting powers were in fact no greater than those of the other two directors. If individual directors had differing views of policy issues, the result could have far-reaching effects on the management of the agency.

Roosevelt's choice for chairman was Arthur E. Morgan, then president of Antioch College; a self-trained engineer who had formed his own engineering company and had designed the famed Miami River flood-control project in Ohio. Morgan was a nationally recognized authority on water-resource development but in addition had evolved a personal philosophy of social and community development that emphasized the values of the small town and integrated resource management. These qualities seemed to make Morgan a natural to try out Roosevelt's ideas about land-use planning, use of marginal land, reforestation, and small-town development in the TVA area.[3]

The other two directors selected by Roosevelt were Dr. Harcourt A. Morgan (no relation to A. E. Morgan) and David E. Lilienthal. Dr. Morgan was then President of the University of Tennessee, a Canadian by birth, who was trained in zoology and entomology and who prior to assuming the university presidency had directed its agricultural programs. David Lilienthal, a young man in his thirties, was a Harvard Law School graduate who had served as Wisconsin State Public Utilities Commissioner. He had completely rewritten Wisconsin's public utility code and gained a reputation as an effective advocate and defender of public power (Figure 2.1).

All three men were dedicated public servants who believed that through TVA they could make major contributions to the public good. All were strong-minded and as it turned out came to different conclusions about the powers conferred on the Board by the TVA Act, and particularly how the President and the Congress had intended TVA to plan for and carry out the overall regional program. In spite of these differences these three board members had to deal with defining the areas of regional concern, the regions to be used in its planning and development programs, and how the agency would relate to other federal agencies and to state and local government within the larger Tennessee Valley region.

A. E. Morgan had spent long hours with newly-elected President Roosevelt discussing the TVA program, and felt he knew the president's wishes concerning the program. But when it came to developing in detail a broad regional development program, the Act itself was not much help and in fact may have raised doubts about what could be done. Did the Act intend TVA to become a corporation responsible only for specific tasks, an agency to coordinate development of the Tennessee Valley, or a new kind of regional government? One observer believes A. E. Morgan eventually took the view that TVA should approximate a regional government in which the agency would assume broad powers and establish and carry out a comprehensive social and economic development program for the Tennessee Valley (Lorenz, 1960) (Figure 2.2).

It seems clear that A. E. Morgan was sympathetic to an integrated planning program for the Valley which would emphasize small communities and rural development. The first project to be planned and started by TVA tends to support this conclusion. At the initial meeting of the TVA Board on June 16, 1933, the first major staff appointment was Earle S. Draper. He was chosen

**Figure 2.1    The original TVA Board. l. to r. Dr. Harcourt Morgan, formerly President of the University of Tennessee; Arthur E. Morgan, Board Chairman, formerly President of Antioch College, a self-trained engineer who designed and built the famed Miami River flood control project in Ohio; and David E. Lilienthal, a Harvard Law School graduate and formerly Wisconsin State Public Utilities Commissioner**

to head the Division of Land Planning and Housing. Draper had been an associate of John Nolen, a city and regional planning consultant, before he opened his own office in Charlotte, North Carolina. He developed a town planning reputation particularly for the planning and building of new towns throughout the South. His best known new town is Chicopee, Georgia, which has been called the ultimate greenbelt city (Aguar, 1995).

Draper's first TVA assignment was to plan a new community near the proposed Norris Dam site. But as it turned out, all three directors did not have the same understanding as to what this assignment was to include. A. E. Morgan had viewed the new town not only as a place to house workers for Norris Dam, but also as a center for training programs for these workers and for upgrading the skills of the people of the region. Morgan had discussed such a program with the President. It is clear from Roosevelt's speeches

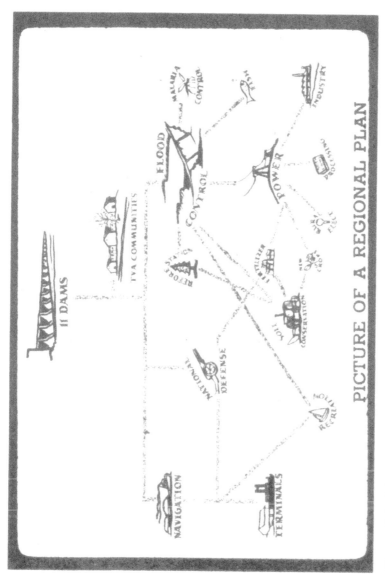

**Figure 2.2  Picture of a Regional Plan. 1934 schematic plan conception by TVA staff showing integration of TVA regional development activities.**

*Source*: Files of J.J. Gray

and writings that he viewed the development of planned rural villages as an essential part of a larger regional planning effort. In addition, it seems also clear that in the discussions between the President and Morgan, the town of Norris was to be the first of a series of planned rural communities, which would be built throughout the Tennessee Valley region to provide places where rural people could learn new skills suitable for use in industrial society (Draper, 1983). Roosevelt even directed the Division of Subsistence Homesteads in the Department of Interior to cooperate with TVA on the Norris town project by providing the additional funds needed to change the dam construction camp, as originally proposed by the Corps of Engineers, to a permanent community (Figure 2.4).

The plans for the Town of Norris proceeded rapidly. Draper quickly followed up on a decision of the Board to buy land for the town by making specific recommendations on site acquisition. In a letter dated August 8, 1933 to J. W. Bradner, Jr., who had been appointed coordinator for the construction of the community, Draper recommended that 2,500 acres be acquired for the town. He noted the need for considerable open space '. . . . for protection, agriculture, and to guard against nuisances in the immediate vicinity of the town.' He also recommended the acquisition of all the wooded area between the town site proper and the dam so as to provide for public ownership of the land between the town and the reservoir shore (Figure 2.5).

The TVA Board officially named the town Norris at its meeting in October 24, 1933. Also to move the project as rapidly as possible Tracy B. Augur was employed on November 2, 1933 and was given the major responsibility for planning the town. Augur was a member of the Regional Planning Association of America which included within its membership such leaders of planning thought in the United States as Clarence Stein, Henry Wright, Lewis Mumford, and Benton MacKaye. Through this group Augur had been a part of the ongoing discussion of city and regional planning and in discussions of urban problems had considered the implications of the British new town movement to city planning and national urban policy in the United States. Augur was also familiar with the concepts Stein and Wright had applied to the new town of Radburn, New Jersey.

The influence on Augur's thinking of the English new towns and the Radburn plan is clearly evident in his account of the planning of the town of Norris in the April 1936 issue of the *American Architect*. In this article he refers to these experiences and to the writings of Sir Raymond Unwin and to the Garden City principles outlined by Sir Ebenezer Howard.

Draper, in his discussion of the purposes of the new town of Norris, also seemed to be influenced by the British new town experience, but he gave special emphasis to the relationship of the new town to agriculture and to the industrial needs of the Tennessee Valley region.

In the planning of the new town . . . which will be accessible to the new Norris Dam, there are certain definite principles which will govern the planning. First, secure a convenient area protected from encroachment by a surrounding belt of land used for agriculture and forestry. Second, to provide for a town of limited size, so planned that agriculture and industry will be integrated to provide a well-rounded economic and social life for the people of the town. Third, to guarantee the protection of the values but that this gain be shared by the entire community rather than by occasional individuals whose pieces of property will be strategically located. Fourth, to plan properly for the best location and arrangement of road, houses, sanitation, education, and recreation (unpublished news release in files at TVA Library, 31 October 1933).

The initial plans for the town reflected a concern for both agricultural training and subsistence farming. The original Town Center included an agriculture building 'intended as a food market which would be supplied, at least in part, by local farmers who would come in with produce and sell from their wagons in the rear of the building'. At this time the thinking was that inhabitants of Norris would engage in part-time farming and the building would serve as a store for selling farm implements, seeds, and farm items. To

Figure 2.3   l. to r. Earle S. Draper, Head of Land Planning and Housing
Division, later redesignated Regional Studies Division,
Engineer Theodore Parker; Harry Tour, TVA Architect;
Roland Wank, Chief TVA Architect. Tour succeeded Wank

*Source:* TVA, 1939.

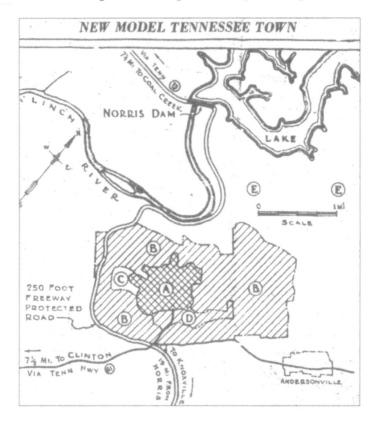

**Figure 2.4   Town of Norris and Environs, TVA Plan from March
1934. (A) central building area; (B) protective greenbelt of
gardens, pastures, and forest; (C) construction unit; (D)
demonstration farm; (E) town forest. Influence of British
New Towns and Radburn, New Jersey is clearly evident**

*Source: Nashville Banner*, 30 March 1934; TVA files.

support this concept the open spaces surrounding the town were to be made
available in four and five acre allotments for subsistence.

But this plan was not to be. The Division of Subsistence Homesteads
withdrew from the project and Lilienthal voiced his opposition to the project
in a sharply worded memorandum to Chairman Morgan. Morgan replied,
defending the project in equally sharp language. The harsh language of these
memoranda suggests that less than six months after the passage of the TVA
Act the differing views of A. E. Morgan and Lilienthal on the scope of the
regional program were already surfacing and becoming a problem in TVA's
program development. These events not only prevented the continuation of
the region-wide rural community building program but changed the purpose

of the Town of Norris so that TVA could provide funds for its completion; it became a demonstration of new town planning and of the effective use of electricity in an urban community (Lilienthal, Morgan, 1933; Morgan, 1974; Gray, 1974; 1–25).

By the last weeks of November 1933, public statements by TVA officials indicated a shift in project justification. Draper stated that it was necessary to provide housing for 2,000 workers building Norris Dam. But at the same time he also raisied the question: why build houses, road, sewer and water supply systems, and electric facilities only to abandon them when they are finished? This became the justification for building a permanent town in the official project report. That report stated that 'In view of the investment that would be necessary in utilities and buildings, it was thought that it would be worth while to go a step further and make the housing of good quality to have a normal useful life instead of building temporary structures'.

The change in purpose also gave Augur and the town planning staff an opportunity to inject into the plans for Norris some of the new town concepts being discussed by city planning professionals At this point the strong influence of the British Garden City movement became very clear. The overall town plan had as central features a green belt completely surrounding the town, the routing of through traffic around the town, in the rolling hills of the town site the roads generally followed the valleys. While Norris shows a kinship with Radburn in the path system and the underpasses to separate pedestrian and vehicular traffic its kinship with the British new towns and the New Deal Greenbelt towns is much stronger. This is evident in the green belt, the common ownership of land, the siting of houses on the best house sites rather than lots, the open space, and the design to emphasize walking rather than in town use of the automobile. Following Stein's ideas on shared amenities, the Norris planners also built group garages for automobiles; but like the experiences elsewhere this concept failed. Town residents preferred having their cars near their home although their boats were frequently stored in the garages (Figure 2.5, Plan of the Town of Norris).

To harmonize the town and house design with the architecture of the region and to relate housing to what was customary and acceptable to the people of the area, members of the new TVA Architectural Staff toured parts of Tennessee, Kentucky, Alabama, and North Carolina 'to take snapshots and measure up houses and to get thorough information as to customs, living standards, etc. which should have an important bearing – in fact form the basis for . . . house design'. This was an important step because most of the TVA architects were from northern states and Draper was determined to create a community that would be acceptable to local people.

The building of the town is an amazing story of people working from a concept and placing houses and streets in locations which fitted the land – all before plans for them had been completed. Individual house sites were located without a regular block and lots system on sites which made use of the naturally rolling topography. This procedure, as Dahir observed in his

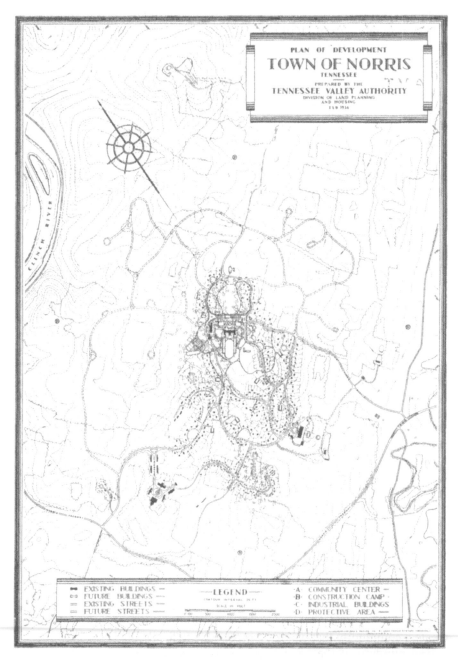

**Figure 2.5   Plan of Development, Town of Norris, Tennessee Valley Authority, Division of Land Planning and Housing, February 1936**

book, *Communities for Better Living*, resulted in a sensitive treatment of the site and gave an unusually natural and informal appearance to the town. The procedure would also result in serious problems in land subdivision when the town was sold to a private developer in 1948 (Dahir, 1950).

Construction of the town moved ahead during the winter, spring and summer of 1934. In January 1934 construction started on the first 150 houses. Earle Draper in April 24, 1984 when interviewed for the TVA Oral History Collection gave credit to Charles Barber, a Knoxville architect for being able to move the project ahead so rapidly. It was Barber's knowledge of materials, dealers, and contractors in the Knoxville area together with his knowledge of house designs that would fit into the informal town plan that made him a key person in the project. Remarkably, by the Spring of 1935 the town as authorized by the Board was essentially complete with roads, utilities, civic buildings and houses in place. The town contained 344 dwelling units – 294 single- family houses, 20 units in 10 duplex units and 30 units in five apartment buildings (Figure 2.6).

Although the town was built in a little more than a year it was a successful demonstration in many ways. First, the small, compact, well-insulated, all-electric homes proved that electric energy could be used for economic heating, cooling, cooking, and lighting. Second, it demonstrated good town design. Tracy Augur in his 1936 *American Architect* article noted that 'Norris is the first self-contained new town in this country to utilize [green belt principles] completely' (Augur, 1936). Finally and probably most important of all it demonstrated that a small community can provide a wide range of urban services and at the same time be a place where people enjoy living. The school became a community school serving not only the students but the adults in the town. There were woodworking and other classes to serve the adults. There were community dances and many other social activities which made the town a good place to live. These demonstrations of Norris to the region and to the nation are recognized by its designation as a National Historic Landmark.

In spite of the Norris success the majority of the TVA Board would not approve the building of new communities at other dam sites. Moreover the TVA Board still faced the responsibility of giving direction to the vaguely defined regional planning and development authority in the TVA Act. In addition, the manner in which the Board organized itself did not help reduce the conflicts. A. E. Morgan's broad interpretation of the TVA Act, particularly his ideas for social and economic planning, alarmed his fellow directors. At a hostile Board meeting in Knoxville on 5 August 1933, Lilienthal voted with H. A. Morgan to approve H. A.'s proposal to divide the supervision of the tasks of the agency among the three directors. Under this plan Lilienthal assumed responsibility for power and legal matters; H. A. Morgan, agriculture, forestry, conservation, and 'rural life' programs; with Chairman A. E. Morgan having engineering, water-resource development, and the regional development program. This plan did not address the question of how

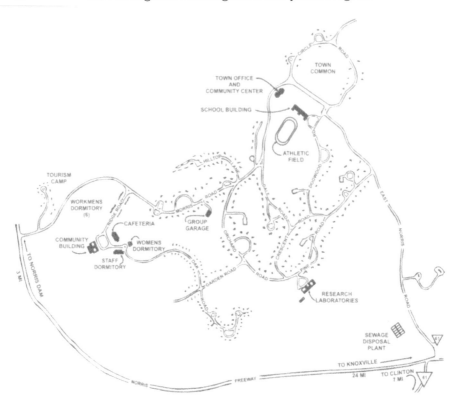

**Figure 2.6   Norris as a Construction Village**

*Source*: TVA files.

overall policies were to be formulated and considered by the Board (Figure 2.7, TVA Responsibilities of Individual Board Members, 1933).

The fact that the Act itself defined specific problems which were regional in scope – navigation, flood control, and power generation – did not encourage a pragmatist such as Lilienthal to give a broad interpretation to Sections 22 and 23. This would have required a definition of what additional problems were in fact regional in scope and which could be administered and implemented at the regional level. Such an approach would also have forced a definition of regional problems as distinct from those which could be handled at the State and local governmental levels. Here again the generalities of Sections 22 and 23 did not help the board overcome the philosophic differences of the three board members.

Soon after the new Board's first meeting on June 16, 1933, TVA began moving ahead with its river improvement program. The earlier surveys by the Corps of Engineers provided an abundance of data on the water resources of

**Figure 2.7  TVA Responsibilities of Individual Board Members, 1933**

*Source:*  National Resources Committee, Regional Factors in National Planning (Washington: US Superintendent of Document, 1935).

the Tennessee River Basin. It also provided alternative plans for development of the river – the high dam vs. low dam systems for improving the Tennessee River. Morgan pressed ahead with these studies and also with plans for the construction of Cove Creek Dam (later to be called Norris Dam) on the Clinch River which had been specifically authorized by the TVA Act. The plan for river improvement was completed in 1935 and was outlined in the March 1936 TVA report to Congress, entitled The Unified Development of the Tennessee River System (see Figure 2.8, TVA Plan for Location of Dams and Reservoirs). The plan chosen was the high dam system, as providing the maximum benefits to the region for an integrated program of navigation, flood control, and power generation.

A. E. Morgan, despite the demurrings of his fellow directors, pressed on with his efforts to establish a program of regional studies and demonstrations. By 1934 Draper, as head of the Division of Land Planning and Housing, had assembled a staff and consultants which included such experienced city and regional planners as Tracy Augur, Ladislas Segoe and Benton MacKaye. Augur was head of the Regional Planning Section and Benton MacKaye had the title of 'Regional Planner.' MacKaye, a New Englander, had been trained as a forester and used his understanding of natural processes as an overall concept for land use planning (Figure 2.9).

Early in 1934 Draper asked MacKaye to outline his views on how the regional planning process should evolve and be carried out by TVA. The record shows that Draper had asked for the Board's approval on February 20 to undertake a project which MacKaye was to call 'The Tennessee Valley Section of the National Plan.' MacKaye's response was dated August 10, 1934 and had as its stated purpose 'to prepare a comprehensive regional plan of the entire Tennessee River Basin, as a guide and co-coordinator of public and private developments now underway or in prospect' (MacKaye 1934). MacKaye's reference to a 'national plan' probably reflected FDR's expressed view that TVA might serve as a prototype for river-basin development across the country, a concept later incorporated in the National Resources Committee publication Regional Factors in National Planning (National Resources Committee, 1935) (see Figure 2.10, Proposed River Valley Authorities, 1935).

MacKaye's memorandum expressed his understanding of the mission of the Division of Land Planning and Housing as including both regional surveys and plan making. The surveys were to include land classification, location of basic resources, and communication and utility facilities. The plan elements would include land use, population patterns, and colonization. These represented some of the traditional regional planning concepts of the time.

MacKaye went on in his August 20 memorandum to note that the regional plan procedure 'seeks to define for the benefit of. . . other [TVA] cooperating divisions (i.e., agriculture, forestry, mineral development, industry, etc.), how the results of their studies might be coordinated with and become a part of

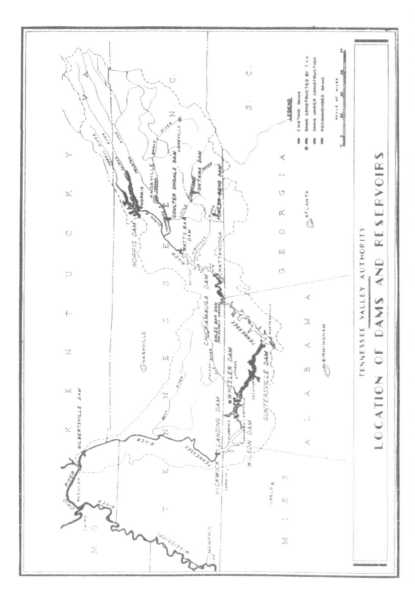

**Figure 2.8 TVA Plan for Location of Dams and Reservoirs, 1936, Tennessee Valley Authority**

*Source:* Personal files of A.J. Gray.

**Figure 2.9    Benton MacKaye in 1964. MacKaye was an innovative TVA
Regional Planner and Forester in the 1930s. While in the
South with TVA he conceived the idea for an Appalachian
Trail extending from Georgia to Maine. With others he
brought the idea to fruition in subsequent decades**

*Source*:    TVA Files.

the physical plan for the Basin.' In a later memorandum entitled 'Physical Planning in the TVA Program,' MacKaye went on to elaborate for Draper his concept of regional planning (MacKaye, 1935). This was an eloquent statement of the potential for TVA to serve as a true regional planning and development agency, as he believed FDR wished it to be. MacKaye called for integrated physical, social and economic planning, the promotion of self-sufficiency for the region, a 'union of plan and action' in which close-range plans would be undertaken within the context of long-range plans, praxis planning – 'we must act while we plan and plan while we act' – and planning as conception rather than coordination.

But such general statements did not help TVA define what and how regional concerns should be included in the plan and how TVA in carrying out the plan would relate to other federal entities and to states and localities which also had a role in regional planning and development. McKaye did not address the question of the limitations on powers granted TVA under Sections 22 and 23 which Augur had described in earlier memoranda. McKaye left TVA soon after, but the ideas he had advanced affected future TVA actions. Parts of his memoranda touched on the problem of coordinating TVA staff efforts in regional planning and development, a problem that was to trouble TVA for many years.

It was also a problem that Draper tried to address shortly after MacKaye left the TVA staff. Draper understood that the Division of Land Planning and Housing was not a central planning staff but was only one of several operating divisions concerned with regional development – one among equals. To deal with the problem, he recommended that TVA establish a Regional Planning Council within the TVA organization. The Council was to aid in coordinated studies and recommendations on TVA's approach to regional problems. TVA divisions having regional development responsibilities were represented on the Council. Draper was secretary to the Council, which was active until he left TVA in 1940.

Perhaps the most significant recommendation concerning the organization of the regional planning and development program came from Dr. H. A. Morgan in a 3 October 1933 memorandum to the other two members of the Board, entitled 'Proposed Statement of Policy in the Planning Activities of the Tennessee Valley Authority' (H.A. Morgan, 1933; see Appendix for full text of this memorandum). To understand the background of this memorandum, it is necessary to recall Dr. Morgan's long association with the land-grant colleges and the county agency system for the delivery of agricultural programs. The memorandum started by noting that TVA had two kinds of functions. The first he described as those 'specific undertakings which are to be executed according to policies set out with some degree of definiteness in law, i.e., dams for flood control, navigation, and the production of power and fertilizer. The second were those broad planning and development functions outlined in Sections 22 and 23.' He recommended that in carrying out these latter functions TVA adopt the policy of conducting surveys and preparing

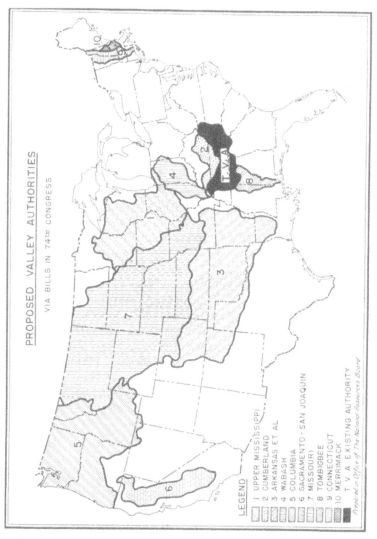

**Figure 2.10    Proposed Valley Authorities in the US, Via Bills in the 74th Congress, 1933**

*Source*: National Resources Committee, Regional factors in National Planning (Washington, DC, 1933, p.106)

plans only in cooperation with existing state and local agencies. Where no agencies existed, TVA would encourage the establishment of such agencies. And, finally, he urged that TVA employ only sufficient staff in these fields to coordinate with and assist these agencies. While his memorandum was not acted upon immediately, the recommendations were to become central to the TVA regional planning and development effort as it evolved.

In addition, this was a compromise which all of the board members seemed willing to accept, particularly since it was a process and did not deal with the question of program content. As it turned out program content probably was determined more by the kinds of technical staffs which made up the TVA organization than it was a result of vague ideological debates within the Board.

Very early TVA found its internal organization, in which individual TVA board members supervised specific staff operations, to be inadequate to the regional development task. (Figure 2.11) In 1934 the Board appointed John Blandford as Coordinator. By an overall reorganization in 1937, Blandford was named General Manager and the Board decided to confine its concerns to overall TVA policy matters. The TVA organization as adopted in 1937 was to remain essentially the same for the next fifteen years. Figure 2.12 illustrates this reorganization and shows the general staff arrangement for dealing with both water improvement and the regional development programs; the latter includes the Regional Planning Council as recommended by Draper.

Although the debate within the board over the scope and content of the regional development program was to continue until A. E. Morgan left TVA in 1938, other specific programs related to regional development began to emerge. Many of these were influenced by the 1936 TVA report to the Congress, which emphasized the relationship between water control in the river channel and water control on the land (Tennessee Valley Authority, 1936). The report noted that in order to control erosion and excessive run-off from the land, TVA programs were already under way in cooperation with other federal and state agencies to control soil erosion, a major problem in the Valley, and to check excessive water run-off from the land. The report also noted the relationship of the reservoir system to economic development through improved transportation and recreation. This, in fact, was the beginning of another key element in the regional development program – the relationship of land development activities to the water improvement program. This question was to become more and more important as TVA became involved in litigation over TVA's authority to distribute power directly throughout the Valley.

The scope of the power program was itself an issue which produced major disagreements between A. E. Morgan and Lilienthal. The power program which finally emerged was to have a major influence on the kind of agency TVA was to become. Lilienthal, with his background in law and utility regulation, contended that TVA could not negotiate with the existing power companies and wanted either to acquire company transmission lines or

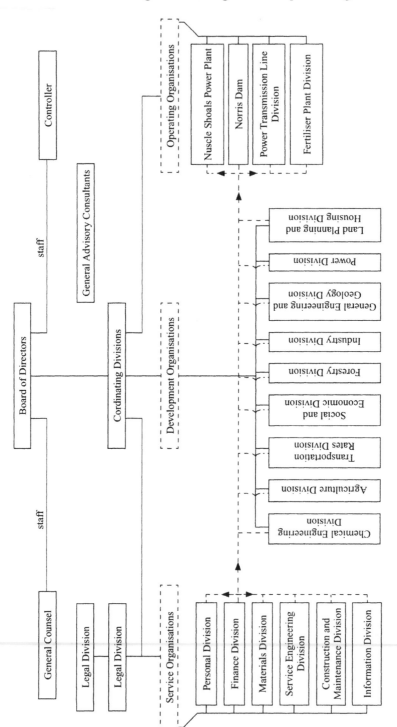

**Figure 2.11  TVA Organization Chart, 1933**

*Source:*   Files of A.J. Gray.

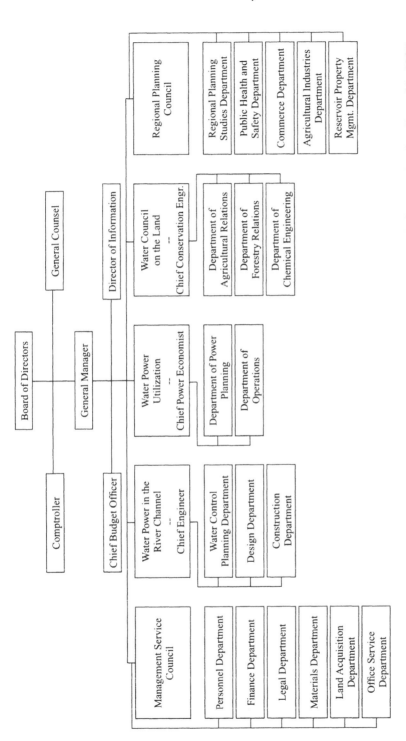

**Figure 2.12　Reorganization of TVA, 1937: Organization Chart of the Tennessee Valley Authority, June 30, 1937**

*Source:*　Personal files of A.J. Gray.

build competing and duplicating lines. Chairman Morgan took a much more moderate view – he believed it was possible to work with the private utilities to achieve TVA's goal of reducing the cost of electricity. Moreover, he did not believe it was necessary for TVA itself to serve directly the whole Valley, preferring instead to have a scientific evaluation of the costs of supplying electric energy to a representative rural-urban area. Such a demonstration would help to determine the true costs of electric power and thus have a basis for a 'yardstick' for government to regulate the utilities (McCraw, 1978, 55).

It is interesting, of course, to speculate what might have happened to TVA if the Arthur Morgan view had prevailed. Certainly the dominant role of power in overall TVA activities would have been diminished and more emphasis would have been placed on the regional development program. However, there was a real question as to whether the regional development program could have survived, in light of the changing political environment, without a strong and dominant operating program.

Ultimately, the Lilienthal view prevailed. Litigation brought by private power companies over the regional monopoly of the power program threatened the very existence of TVA. As a result, both the Board and staff took a more restrictive view of the content and management of the regional development program. The lawsuits brought by private power interests against aspects of the TVA power program also strengthened the Lilienthal position that TVA should concentrate on the specific activities spelled out in the TVA Act.

Draper, speaking at the City of Norris Fiftieth Anniversary Celebration in 1983, recalled a conference he had with Lawrence Fly, TVA General Counsel, in 1936, after the narrow (five-four) Ashwander decision of the U.S. Supreme Court had permitted TVA to go ahead with direct power sales to municipalities and cooperatives.[4] (The Ashwander case had challenged TVA's right to sell or distribute electric power generated at Wilson Dam or any other federal project in the Tennessee River Basin.) Had this decision gone the other way, the power part of the TVA program as devised by Lilienthal would have been eliminated. One thing is certain: the close court decision pushed TVA to take a defensive position in so far as the regional development program was concerned. There had always been concern over the meaning of Sections 22 and 23, and Draper noted that Fly took the position that work under these sections should be related directly to the river improvement program because he did not want to have to defend work done under these sections on a broader basis (Draper 1983).

It was not only the litigation which was a major TVA concern. The constant clamor for restricting TVA powers and activities at times created almost a paranoia among top TVA officials. For example, in the May 5, 1945 issue of *Business Week* the Tennessee Industrial Planning Council, an advisory body to the Tennessee State Planning Commission, purchased an ad featuring the advantages of 'moving your plant to Tennessee' with one of these advantages low cost electric power. Shortly after the ad appeared Chairman Lilienthal wrote to P. D. Huston, Chairman of the Tennessee State Planning Commission questioning the approach that to obtain new development 'it is necessary to close plants in one part of the country in order to establish new plants in another'. This reflected TVA's fear of antagonizing people in other parts of the country; people whose support TVA felt was necessary to continue the TVA power program.

Another little-recognized but important decision was that relating to the use of force-account construction of its major projects. Most other federal projects relied on contract construction, but this would have resulted in long delays in preparing detailed specifications and advertising of bids. Thus TVA decided to use its own staff to design and build its projects (Durisch and Lowry, 1953; 220). This decision was important to the regional development program because it permitted TVA to broaden its in-house staff skills, which would be available to deal with problems arising out of the river improvement program, and to relate that program to the development of the region.

In spite of the TVA's restrictive view of Sections 22 and 23, it did understand that the river improvement program of the scale proposed would have a massive effect on the physical, social, and economic development of the watershed and beyond that to the seven valley state region.. As a result, it took a very broad view of the relationships between the river improvement and regional development programs. In addition, an important by-product of relating these two aspects of the TVA program was that budget requests for the regional development staff could be more easily justified. This became a dominant consideration as the nation became less and less supportive of the New Deal and Roosevelt's attention was diverted to other problems facing the nation.

## The Nature of the TVA Development Plan and Program

Some critics, such as John Friedmann, have argued that TVA lacked a development plan (Friedmann 1955). But in its early years TVA developed and closely followed an explicit, discrete physical plan – the plan for the integrated development of water resources of the Tennessee River as transmitted to the Congress in 1936 (Figure 2.8). These were certainly appropriate elements to include in a plan for a watershed region.

This plan was supplemented by a series of policies designed to provide a guide for the regional program carried on in cooperation with the seven valley states. These polices also helped to relate the TVA's regional development activities to its river improvement program. One central policy, as noted above, had its origins in the October 3, 1933 memorandum from Dr. H. A. Morgan to A. E. Morgan and David Lilienthal, which recommended that all surveys and studies undertaken by TVA under Section 22 be carried out in cooperation with appropriate state and local agencies, and that where no agencies existed TVA would stimulate and promote the establishment of such agencies. This set the tone of the regional development program as essentially one of regional institution building.

In practice, state agencies were the key agencies in the regional development program and those with which TVA established its primary relationship. A related policy provided that surveys and programs, wherever possible, should be designed to serve an entire state rather than serving only the area within the Tennessee River drainage basin and the TVA power service area. This broadened the Tennessee Valley Region to include the seven Tennessee Valley states, parts of which were located in the Tennessee River Basin (Figure 2.13, map of the TVA Planning Regions and the TVA Power Service Area).

Both the TVA Board and staff understood that to have viable state and local programs in the watershed and power service area it was necessary that these programs operate and have state-wide support. They also understood that for planning and plan-making to be meaningful they must be related to the decision-making process of those implementing the plans so that the plans could be made a part of operations (Birkhead, 1962; 87–93).

In order to provide some parameters to the regional development program, it was the general TVA policy to encourage state and local activities that had a reasonable relationship to the improvement of the river. To the credit of the TVA Board, the kinds of activities to be encouraged were viewed very broadly, with the result that both the scope and the area covered by these programs extended beyond the immediate needs for river improvement and far beyond the area included in the Tennessee River Basin.

These policies, plus the plan for river improvement, constituted TVA's regional development plan. Although not in the tradition of accepted city and regional planning practice of the time, which emphasized discrete land use plans to be implemented by regulatory and advisory actions, it was a plan that was feasible, one that met the needs of a developing region, and one that built institutions capable of helping to develop an area larger than the Tennessee watershed.

By means of this plan, TVA found a way to deal with the special kinds of problems that regional agencies have in plan preparation and implementation; i.e., TVA supplemented its powers, authority, and financial resources with those of state and local governments. In addition, it recognized that ours is a government of shared powers (federal, state, local). It provided a guide for determining the problems that were uniquely regional and problems that were shared with other governmental units. It recognized that different regional definitions were required to address different regional concerns and problems. And it outlined how TVA, as a regional agency, could relate to traditional governmental units.

This view of the TVA development plan and program is not shared by everyone. Philip Selznick in the 1966 edition of his book, *TVA and the Grass Roots*, describes the TVA regional development program in this way:

My conclusion was not merely that TVA trimmed its sails in the face of hostile pressure. More important is the fact that a right wing was built inside the TVA.

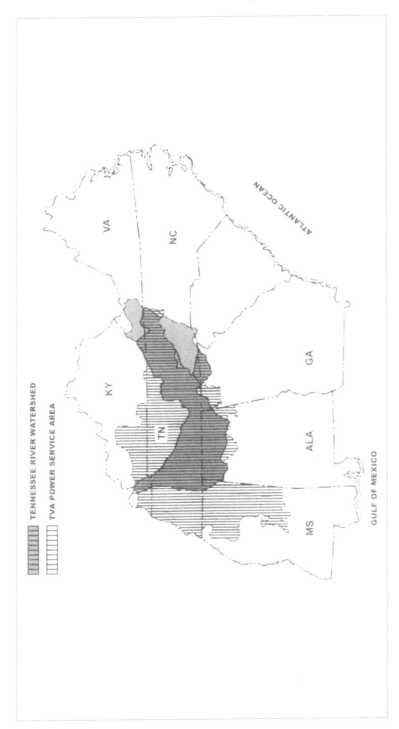

**Figure 2.13  TVA Planning Regions, Drainage Basin, Power Service Area, and Tennessee Valley States**

*Source:*  Tennessee Valley Authority, 1991.

The agricultural program of the agency simply turned power over to a group that had strong commitments, not to a distinct ideology but to a specific constituency. This group then became a dynamic force within TVA, able to affect programs marginal to the agricultural responsibilities of the agency but significant for conservation and rural life.

This was not a case of simple compromise made by an organization capable of retaining its internal unity. Rather, a split in the character of the agency was created. As a result, the TVA was unable to retain control over the course of the basic compromise. Concessions were demanded and won which may not have been essential if there had been fundamental unity within the agency' (Selznick, 1949, xiii).

Reviewing Selznick again at this late date suggests that although his theoretical conclusions were reasonable he did not grasp either the nature of the internal TVA organization or the basic conflicts built into the agency by the attempts in the TVA Act to merge the concepts of integrated water resource development and the then new and emerging ideas of regional planning. Neither does he spell out the actual alternatives open to TVA given the composition of the original TVA Board and the vague guidelines for regional planning and development in Sections 22 and 23 of the Act. If Tracy Augur's conclusions concerning these sections are correct, and we tend to agree with them, TVA did not have any significant developmental powers beyond its river responsibilities.

It is true that the agricultural staff tried to influence TVA programs marginal to its responsibilities; but the same can be said for all TVA professional staffs – the foresters, the fish and wildlife experts, the public health people, the commerce and industrial development people, the regional planners, the lawyers, the engineers, the power people, and the land acquisition people. Certainly the regional planning staff had a greater impact on reservoir shoreline purchases than the size of the staff would seem to warrant.

Selznick relies heavily on Lilienthal, quoting the latter's book, *TVA: Democracy on the March*, almost to the exclusion of other sources. (Lilienthal, 1945) There is, for example, not a single reference to the two Morgans in his list of sources. Nothing is said of H. A. Morgan's concept of what he called the 'Common Mooring' – the common cultural, traditions, and beliefs of the people, that in Morgan's view had to be taken into account in TVA's operations. Nor is any reference made to H.A. Morgan's long association with land grant colleges and the county agent system of the delivery of services to farms. In addition, no mention is made of the influence of the teachings of Edward Bellamy on A. E. Morgan's personal philosophy of social and community development which emphasized the values of the small town and integrated resource development. Lilienthal was a lawyer and in the formative years of the TVA program took a very narrow view of the powers granted to TVA. He only began to promote the grass roots philosophy after he became chairman and by that time the scope and content of the regional program had been established.

It also is important to recall that the concept of the regional development program based on regional institutional building and TVA cooperation with existing and new institutions emerged within six months of the passage of the TVA Act. During that period the Board was struggling to find a concept of 'regional resource development' which was acceptable to all board members. As noted earlier it was the October 3, 1933 memorandum written by H. A. Morgan to A. E. Morgan and David Lilienthal entitled 'Proposed Statement of Policy in Planning Activities of the Tennessee Valley Authority' which set the tone for the organization of TVA's regional planning and development activities. It is not surprising given H. A. Morgan's background in the workings of the land grant colleges that he was the one to make this proposal. Too, it fitted well with Lilienthal's view that TVA should concentrate on the specific powers granted to it by the Act.

It was also H. A. Morgan who proposed a specific program for using the World War I nitrate plants at Muscle Shoals, Alabama. Reasoning that nitrogen was a renewable resource which could be replaced through good agricultural practices, Morgan proposed that TVA concentrate on the research and production of phosphate (a non-renewable resource) for use in a new breed of fertilizers. The program was to have three parts: research into new, less expensive, but more effective fertilizers; testing the new fertilizers under actual farm conditions; and working with fertilizer manufacturers to assure the proven fertilizers were available commercially.

Given H. A. Morgan's long association with the agricultural departments of land grant colleges it was natural that he would recommend turning to this source for help in the research and testing aspects of the program although its should be noted that most of the research was done by TVA's own staff at the Muscle Shoals plants. Implementing the program through the land grant colleges would seem to Morgan the only logical approach to programs designed to deal with agricultural problems.

TVA began its agricultural program in 1933 before other large national programs such as that of the Soil Conservation Service were made operational. True, it might have been possible for TVA to set up a new internal organization to meet its needs; one that would compete with the agricultural programs of the land grant colleges but H. A. Morgan and Lilienthal were dedicated to keeping small staffs for the regional development program . By working with the land grant colleges and the county agents TVA was able to have the colleges modify their programs to meet TVA's need to test its experimental fertilizers. This was an application of the staffing and operational policies proposed by H. A. Morgan in his October 3, 1933 memorandum; policies that provided a compromise which both A. E. Morgan and David Lilienthal could support as the board tried to carry out its specific duties for flood control, navigation and power production and at the same time organize the overall regional development program.

It seems to us that TVA's reservoir land purchase and use policies are the central feature of Selznick's argument that the agricultural 'group became a dynamic force within TVA, able to affect programs marginal to the agricultural responsibilities of the agency'. But as has been noted there were many TVA staffs that significantly influenced reservoir land policies. Individual staff views were affected by internal and external forces as well as by the perceived effectiveness of the land policies by the TVA Board. It is true that the agricultural staff generally opposed extensive fee purchases of land by TVA for reservoirs or shoreline development. But other TVA staffs took similar positions in many cases. In any large organization made up of a large number of highly trained and specialized professionals one would expect differing staff views on such a new approach of reservoir land acquisition. Up until this time the normal practice had been to acquire only the land actually flooded by the reservoir and in most cases the practice was to acquire only easements.

Within TVA the chief advocate for purchase of reservoir shoreland was the Regional Planning Staff. This staff believed that the purchase of a narrow strip along the reservoir edge was not sufficient. It viewed the reservoirs as adding new resource combinations to the region by exchanging agricultural land for new recreation, transportation, and industrial resources and that these resources should be used to support the changing demographic, social, and economic needs of the Valley region.

The Regional Planning Staff worked with state and local planning agencies and with other TVA staff to locate reservoir shoreland suitable for state and local parks, wild life refuges, industrial development, marines, and shoreline residential use. It is true the agricultural group resisted such acquisitions and did gain some concessions on certain reservoirs. Eventually TVA developed policies for the sale and lease of TVA-owned lands to public agencies and to private individuals on the basis of proposals which would carry out reservoir land use objectives.

Outside forces frequently were more important in shaping TVA land purchase and disposal policies. Congress became concerned with government ownership of large amounts of land. For example, the House Appropriations Committee at its 1948 TVA budget hearing directed TVA to present a plan either to make the Town of Norris and the Wilson Dam Villages (built during World War I) self supporting or to dispose of them. The towns were sold in 1949. Again, in 1953, the General Services Administration, under the policies of President Eisenhower, directed TVA to review all its land holdings and to dispose of any surplus properties.

These pressures brought about significant changes in TVA land use policy. Whereas formerly a specific proposal designed to carry out reservoir land use objectives was required as a condition of TVA land disposal; now TVA staffs were required to provide a justification for land retention by TVA. As a result, thousands of acres of land were sold and most of these sales did little to promote regional and reservoir development. It was not until the late 1950s that this policy was reversed and pre-impoundment reservoir shoreland planning became the norm. It was then that the shoreland planning for Melton Hill, Nickajack, and Tellico reservoirs was undertaken.

In summary Selznick is correct in asserting that the agricultural group was an important influence on TVA policy. But it was not the only influence and in our opinion was not the most important influence.

## The Scope of the Regional Development Program

In the last thirty years a debate has raged over the statistics that measure how well the Tennessee River watershed and TVA power service area have fared economically as compared to other parts of the Southeast and the Southeast as a whole. (Jacobs, 1984; Pulsipher, 1984) In this debate there was always a question as to which statistics and areas were comparable. The pros and cons of this debate have been well documented in many books and articles. We will not add to this debate. Instead we will concentrate on the projects and activities undertaken as part of the regional development program and how they related to the needs of a changing region.

It is the contention of the authors that it was TVA's emphasis on institution building in the seven Valley states that helped the Tennessee Valley and the Southeast to move as rapidly as they did in their transition from an agricultural to an urban-industrial economy. It is important to recall that in 1930 the South was still an agricultural and resource-based economy with farms in much of the area operating at a bare subsistence level. Roosevelt had characterized the South as the nation's number one problem. In 1930 half of the population lived and worked on farms. In the twenty years that followed the farm population declined by nearly 1.6 million or a third of the total farm population. During the same period the urban population grew by 48 percent and the non-farm population grew by 75 percent. These rates of growth were faster than those of the nation as a whole. By 1970 the population distribution of the South approximated the national norm (US Bureau of the Census, 1930, 1940, 1950).

In the twenty years between 1933 and 1953, the Tennessee Valley and the Southeast were well on their way in this transition. The patterns of population distribution (urban, non-farm outside urban areas, and farm), employment and income in both areas had already started to mirror national patterns. The authors do not contend that this change was solely the result of TVA's programs but rather that the combined effort by TVA and the seven Valley states helped to ease the transition within the Valley and the Southeast from a farm to an urban industrial economy.

TVA had been able to move rapidly on the program of integrated water-resource development for the Tennessee River thanks to the availability of earlier Corps of Engineers studies and plans for the development of these resources and because the Bureau of Reclamation had already started on construction plans for what were to become Norris and Wheeler dams. Norris and Wheeler dams were begun in 1933; these were followed by Pickwick Landing Dam in 1934, Guntersville Dam in 1935 and Chicamauga and Hiwassee Dams in 1936. Thus in the short span of three years after its creation TVA had been able to make rapid progress in carrying out its plan for improvement of the Tennessee River as submitted to the Congress in 1936 (see Table 2.1).

Table 2.1 shows the extent to which TVA made purchases above normal reservoir pool levels. There are substantial variations from reservoir to reservoir in the amount of overpurchases as a result of differences in available resources and development opportunities. The largest relative overpurchases were at Fontana Reservoir where large amounts of land were purchased to reduce the need for costly road relocations; land that was subsequently turned over to the National Park Service so that the Great Smoky Mountains National Park could be extended to its proposed southern boundary. Large overpurchases were also made at Norris reservoir for the same reasons, lands which were later made available to the state of Tennessee for parks and wildlife protection areas.

Both A.E. Morgan and Dr. H.A. Morgan, because of their professional backgrounds and experience, understood that control of water in the river channel would be insufficient given the then current agricultural and forestry practices in the Valley, which resulted in both heavy water run-off and soil erosion. To both Morgans 'water control on the land' was an obvious and necessary complement to 'water control in the channel'. Thus early parts of the regional development program concentrated on problems of agriculture and forestry. The staff responsible for these programs were for the most part trained in the natural sciences: botanists, biologists, zoologists, wildlife specialists, and soil scientists.

The natural scientists contributed most to the early TVA agenda. But it was the breadth of the overall in-house TVA professional staff that extended the regional program beyond concern with natural resources. It was A. E. Morgan who had recruited the talented Earle S. Draper to head the Land Planning and Housing Division. Draper in turn recruited city and regional planners, architects, landscape architects and geographers. The TVA staff also included a social and economics group, also most likely recruited by A. E. Morgan, which was made up of economists, sociologists and political scientists. Because of his interest in education Morgan hired educators who helped to develop the employee training programs that became the hallmark of TVA construction projects.

This wide-ranging staff of planners and social scientists was shortly to be supplemented by people concerned with economic development. Under the influence of this staff the TVA regional program included natural resource development plus the development of facilities and services important to the region as it moved from an agricultural to an urban-industrial economy. Some of the key programs – agriculture, forestry, fish and wildlife, and regional planning and development – were the foundation of the early TVA regional development program and merit a brief review.

**Table 2.1  TVA Fee Purchases of Reservoir Land; 17 Reservoirs Built, 1933–1953**

| Reservoir and Year Closed | Total Acres required | Acres in Maximum Pool | Acres Above Normal Pool | Percent of Total Acreage above Normal Pool |
|---|---|---|---|---|
| *Mainstream Reservoirs* | | | | |
| Kentucky (1944) | 228,539 | 160,300 | 68,239 | 30% |
| Pickwick (1938) | 63,625 | 43,100 | 20,525 | 32 |
| Wheeler (1936) | 103,486 | 67,100 | 36,386 | 35 |
| Guntersville (1939) | 109,671 | 67,900 | 41,771 | 38 |
| Chickamauga (1940) | 59,600 | 35,400 | 24,200 | 41 |
| Watts Bar (1942) | 50,647 | 39,000 | 11,647 | 23 |
| Total, Mainstream Reservoirs | 615,568 | 412,800 | 202,768 | 33 |
| *Tributary Reservoirs* | | | | |
| Appalachia (1943) | 12,407 | 1,100 | 11,307 | 91% |
| Hiwassee (1940) | 23,911 | 6,090 | 17,821 | 75 |
| Chatuga (1942) | 8,680 | 7,050 | 1,630 | 19 |
| Nottely (1942) | 5,958 | 4,180 | 1,778 | 30 |
| Norris (1936) | 144,926 | 34,200 | 110,726 | 76 |
| Fontana (1944) | 66,749 | 10,640 | 56,109 | 84 |
| Cherokee (1941) | 43,564 | 30,300 | 13,264 | 30 |
| Ft. P. Henry (1953) | 1,275 | 872 | 403 | 32 |
| S. Holston (1950) | 11,996 | 7,580 | 4,416 | 37 |
| Watauga (1948) | 9,198 | 6,430 | 2,768 | 30 |
| Total, Tributary Reservoirs | 332,466 | 108,922 | 223,544 | 67 |
| Total, All 17 Reservoirs | 948,034 | 521,722 | 426,312 | 45 |

*Note*: Ft. Loudon (1943), Douglas (1943), Ocoee #3 (1942), and Boone (1952) reservoirs were not included because of the special land acquisition policies used.

*Source*: TVA Handbook, July 1972, pp. 56, 387.

*Agriculture*

TVA decided to center its agricultural program around research in superphosphates, demonstration of the use of these fertilizers under actual farm conditions, and to encourage the production of proven fertilizers by private companies. Here a major objective was to encourage the farmers of the region to change from row-crop to grassland agriculture. Dr. H. A. Morgan in making this recommendation reasoned that the growing of legumes would restore nitrates to the soil, and adding phosphates would help improve the generally depleted soils. To do this the Muscle Shoals Nitrate Plants were converted to phosphate research and production even though the plants, by law, also had to be maintained in standby condition to produce nitrates for national defense.

As early as 1934, TVA entered into a memorandum of understanding with Valley state land-grant universities and the U.S. Department of Agriculture for a coordinated program of agricultural research, extension, and land-use planning. In the following year the first test-demonstration farms were chosen in the Valley states. This program was to spread throughout the South to speed the change to soil conserving and more market-oriented agriculture. By 1953 there were nearly thirteen thousand such farms in the Valley and seventeen thousand in the rest of the Southeast and the nation (Allbaugh, 1953). Although no new agricultural institutions were created, TVA clearly helped change the programs of established farm agencies and, through these agencies, changed farm practices in the Southeast and across the nation.

*Forestry*

When TVA was established, over sixty percent of the land in the Tennessee Valley was already in forest use. Forestry was a critical economic factor in the Valley. The TVA Act, moreover, gave special attention to the use of marginal land and to reforestation. It is not surprising that a forestry program was started early, to help control water run-off and prevent erosion and also to assure that the forest resource contributed to the economy of the region. The TVA forestry program is one of the best examples of TVA's effort to establish new institutions and programs to deal with the most urgent forest problems in the Valley and, once established, for TVA to move to other forestry problems that required new approaches for their solution.

In the Tennessee Valley, as elsewhere in the South, it was a common practice to burn timberlands in order that the new growth could provide forage for cattle. TVA, beginning in 1933, directed its first efforts at forest improvement by working to establish a system of fire control in each to the Valley states. By 1953, 80 percent of the forest lands of the Tennessee Valley region had some kind of protection operated and paid for by the States and local agencies. In 1954 Congress approved the Southeast Forest Fire Control

Compact to allow states in the region to enter into agreements for cooperative planning and forest fire prevention.

Over the next ten years the forestry program was to move rapidly on other forest-related problems. Beginning in 1934, TVA started a program of reforestation under which seedlings from TVA nurseries were given to landowners without charge. By 1949 more than 200 million trees had been planted. After 1956 the seedlings were made available through state seedling producing programs. In 1939 TVA started a continuous forest inventory project which would provide information on the quantity and quality of the forest resource, the rates of growth, and provide a base for estimating use rates that would not deplete the forest resource. This was to become a valuable tool for both the states and TVA to gauge the kinds and size of forest industry that could be encouraged to locate in the area.

From 1940 to 1943 TVA worked with Valley states to develop programs in basic forest management not only on the larger individually owned tracts of forest land but also on the large acreage of forest land on existing farms. In the same year TVA started programs with the states to encourage the growth of industries to use the large forest resources of the Valley and the Southeast. In the Tennessee Valley alone there is now a billion-dollar forest-products industry; in the Southeast as a whole it is proportionately larger.

As these programs were gradually transferred to Valley state forestry agencies, TVA initiated a program aimed at upgrading the forests through genetic research. By producing superior trees the forests yields are increased. The results of this work were made available to the forestry industry throughout the nation. The TVA forestry program is another example of institution building that improved conditions not only in the Tennessee Valley but throughout the Southeast and the nation.

*Fish and Wildlife*

TVA reservoirs have over 600,000 acres of water surface and over 38,000 miles of shoreline. This new resource presented unique opportunities for relating the TVA reservoir system to national wildlife and particularly wildfowl management. It was an opportunity to supplement the already existing Mississippi River fly-ways for migrating birds.

The program started in 1939 when TVA transferred roughly 41,000 acres to the Bureau of Sport Fisheries and Wildlife to create the Wheeler National Wildlife Refuge, with three-quarters of the land being used for food production. As a result, thousands of ducks and geese are now attracted to the area.

An additional 120,000 acres have been made available to federal and state agencies for waterfowl management. Many of these areas are in the Kentucky Reservoir areas, which form a natural supplement to the Mississippi fly-ways.

**Figure 2.14  Willis Baker, Director, Forestry Relations Division. Baker was a key figure in developing the forest resources of the Tennessee Valley and the Southeast US**

*Source*:    Patricia Bernard Ezzell, TVA Historian.

*Inter-regional Freight Rates*

As noted above, not all TVA programs were related to land resources. For example, TVA started to look into the effect of disparities in freight rates between the Northeast and the South on development of the southern part of the United States. The freight rate structure in the 1930s tended to perpetuate the then existing regional specializations which penalized the South and West, keeping them in their long-time dependence on raw-materials production, while the Northeast could continue to dominate the manufacturing field without fear of competition from other parts of the country. The problem was that manufacturers in the South could not ship goods to the Northeast, where the major markets existed, at rates comparable the those given to manufacturers from the Northeast to ship their goods to the South. It was a patently unfair situation.

**Figure 2.15  Abraham H. Weibe, first head of the fish and game Branch in the Forestry Division. He identified the new fish and game resources along the TVA reservoir system and also possible fish and game reserves and refuges**

*Source*:  Patricia Bernard Ezzell, TVA Historian.

The first TVA report on the problem was submitted to the President on 28 May 1937. Incidentally, this was to be the first Section 23 report submitted by TVA to the President and the Congress. The report discussed the reasons and remedies for the barriers to interstate commerce created by the then existing freight rate structure. It also noted there was no national rate structure but rather a group of regional rate structures, which hindered development of major parts of the United States, including the South. A second report was sent to the President on 9 February 1939 which reemphasized the need for freight rate restructuring if the Tennessee Valley and the South were to have an equal opportunity to develop.

Finally, TVA provided testimony before the Interstate Commerce Commission at its hearings in 1942. A. D. Spottswood, testifying for TVA, took the position that the 'economic consequence of the regional disparities in existing class-rate structure. . . tend to perpetuate an economic maladjustment which interferes with the orderly development of resources of the South and West' by placing these regions at a substantial disadvantage to market manufactured goods in the Northeast, the region that then contained

**Figure 2.16   Aelred J. Gray, TVA Planner 1935–1973; Chief, Urban
Community Relations Staff, 1945–1965; Chief, Regional
Planning Staff, 1965–1973**

*Source*: Patricia Bernard Ezzell, TVA Historian.

51 percent of the population of the U.S. and represented the largest consumer market in the nation. He continued by recommending the establishment of class-rate levels that would benefit no region (75th Congress 1937; 76th Congress, 1939; Spottswood, 1942).

The rate structures were changed in the 1940s, effecting a major change in institutional policies which were important to the development of the South as well as the Tennessee Valley.

*State and Local Planning*

As noted above, the first major assignment of Draper's Division of Land Planning and Housing was to plan and build the new town of Norris, which was originally conceived as the first of a series of such communities in the Tennessee Valley. When the TVA Board as a whole failed to support this program, Draper and his staff started looking for region-wide programs more in keeping with the emerging policy of regional plan preparation in cooperation with existing state and local agencies, and with activities that could be related to the river improvement projects.

There was a close working relationship between TVA and the National Resources Planning Board. Both were interested in state and local planning and had agreed to work together to stimulate such agencies.

An opportunity came in 1935 when the then Tennessee Governor McAllister created the Tennessee Valley Commission to work with TVA on development problems. Draper employed Alfred Bettman, a prominent planning lawyer from Cincinnati, to work with the commission in the preparation of state, regional and local planning legislation. The Tennessee General Assembly adopted the recommended legislation the same year, legislation which was to become one of the models for such legislation throughout the country.

Several of the Valley states also began to undertake state-wide studies as a basis for overall state planning. For the most part, such studies were concerned with basic natural resources and with economic conditions. TVA helped state agencies with many of these studies by providing both technical assistance and data on natural and economic resources.

As the river improvement program progressed, one of the emerging problems was the need for planning physical and economic adjustments in reservoir-affected communities. Tennessee had organized its State Planning Commission in 1935 and TVA joined with that agency in an effort to find ways to help these communities. The result was a memorandum of agreement between the Tennessee State Planning Commission and TVA, by which TVA would provide funds for Tennessee to employ city and regional planners to assist such communities; the state in turn agreed to organize a state-wide program of local planning assistance to help the reservoir-affected communities as well as urban communities throughout the state. The TVA contributions continued for a few years as Tennessee gradually worked toward full-state funding of the program. In 1935 there were only two or three local planning agencies in the state; by 1946 there were 74 local planning agencies in Tennessee. Similar programs of local planning assistance were organized by all of the Valley States. In fact, the Tennessee Program became the model for the nation and the concept of that program was made nationwide through the inclusion by the Congress of Section 701 of the Housing Act of 1954.

Although a major effort was focused on organizing state planning agencies and local planning assistance programs in each of the Valley States, the TVA regional planning staff also made a concentrated effort to relate the TVA reservoirs to state and local plans and programs. One of the first major recommendations was that TVA find ways to control the reservoir shoreline. Initially, it was agreed that TVA should purchase a minimum strip 250 feet wide around the reservoir. But with the growth of state and local planning as well as recreation and wildlife resource development agencies, more detailed reservoir plans were prepared to relate the new reservoir shoreland resources to state and local development plans. For, example, the Alabama state planning and resource development agencies, the Guntersville City Planning Commission and TVA prepared a plan for the Guntersville region shoreline uses which included a state park, shoreline residential development, public and private marinas , and waterfront industry. Within Guntersville, a detailed plan was prepared and implemented to reconstruct the Guntersville shoreline for recreation, industrial and residential development (see Figure 2.17, Waterfront Park and Harbor Plan for Guntersville, Alabama).

In Decatur, Alabama, state agencies, the Decatur City Planning Commission and TVA prepared a plan which integrated the Wheeler National Wildlife Refuge with city plans for residential, industrial, and park development while a regional plan included a major marina, a planned industrial area, an Alabama State Dock, and shoreline residential development. Similar detailed land use plans were developed and implemented on TVA reservoir and adjacent lands in dozens of other cities and towns along the River (Gray and Roterus, 1960).

*Economic Development*

In the early 30s, very few states had separate agencies concerned with economic development. The state planning agencies tended to fill this void by studies of natural and economic resources and by investigation of industries that could utilize these resources. TVA not only joined states in these studies but also encouraged the states to help identify the lands along the TVA reservoirs, which were suitable for industrial use.

*Association of Valley State Planning and Development Agencies*

By 1945 most of the Tennessee Valley states had state planning agencies or development agencies with state planning functions. Beginning in that

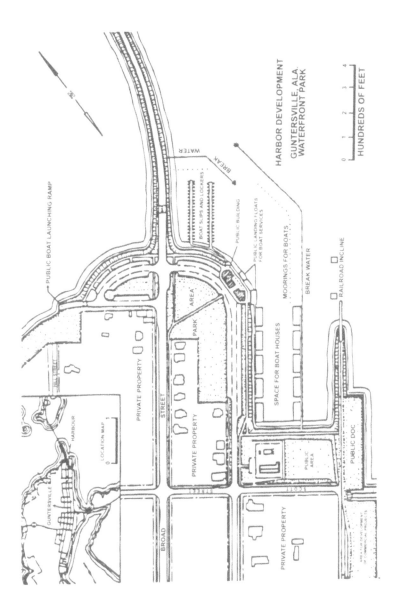

**Figure 2.17 Harbor Development, Guntersville, Alabama, Waterfront Park**

*Source:* Gray and Roterus, 1960.

year annual conferences were organized between the Directors of the Tennessee Valley state planning and development agencies and TVA staffs. The overall purpose of these meetings was to try to expand the cooperative relations between the states and TVA. A basic assumption was that the state planning agencies could identify state programs that could help with valley development problems, help find ways to meet problems created by or emerging from TVA activities, and help to organize new programs that might be needed in the Valley.

Some examples of topics covered in these conferences were the exchange of information on research underway, identifying industrial opportunities in the Valley in both general and specific terms, organization of state-wide local planning agencies, the need for training programs in city and regional planning to meet the personnel shortage in these fields, the recreation potential of TVA lakes, and water-resource data needs.

These conferences were to continue for several years, but the state agencies were interested in meetings involving all the Southeast states. Eventually the Southern Association of State Planning and Development Agencies was organized and TVA worked with that group. These programs in state planning and local planning assistance, originally stimulated by TVA, covered the entire South and helped meet the needs of a changing region and its growing cities.[5]

*Recreation*

The system of TVA reservoirs added a new recreation resource to the Tennessee Valley. Here, too, TVA activities were directed to the development of state-wide recreation programs. In 1934 Draper approved a recreation planning study which identified potential scenic and recreation areas throughout the Valley. The Scenic Resources of the Tennessee Valley, produced by a team headed by staff planner Robert M. Howes, provided an extraordinarily valuable guide to state park and recreation agencies in carrying out their programs to acquire and protect unique lands for the public (see Figure 2.18 for an example of a portion of the scenic resources atlas). As a result, the Valley states today have some of the best state park systems in the nation (Tennessee Valley Authority, 1938). TVA's development approach in recreation and parks was to foster institutional development at the state and local level, and to show the new institutions through demonstration projects how to develop and manage resources.

As a part of its regional demonstration program, TVA built state parks on Norris reservoir, which later became Big Ridge and Cove Lake State Parks. TVA encouraged Tennessee to expand its conservation program to include a state park system. Over the years, Tennessee assumed responsibility for Big Ridge and Cove Lake and expanded the system so that there would be a state park within fifty miles of every citizen in the state.

Other Valley states organized similar state park systems. These parks were important to the development of the Southeast as well as the Tennessee Valley. There are now seventeen state parks on TVA lakes.

The states also helped encourage local park systems as a part of the growing urban systems. TVA reservoirs and shorelands are the sites of over eighty municipal and county parks as well as 423 smaller public access and roadside parks operated by federal, state and local agencies.

## Professional Training

The many skills represented on the TVA staff resulted in strong ties to national and Southeastern professional organizations. As the regional institutions using many of these same skills grew, so did the demand for professionals in these fields. TVA professionals supported and worked with many regional colleges and universities in an effort to meet the demand. Two examples will be used to illustrate the effects of TVA support for professional training in the South.

*City and regional planning* The growing number of planning agencies in the South resulted in a demand for professionals trained in city and regional planning. None of the Southern universities had training programs in this field. The University of North Carolina at Chapel Hill proposed to start such a school and asked for TVA's help. TVA provided funds to support one faculty member and in addition provided staff as visiting lecturers and offered internships to students from the school. This program provided professionals throughout the South and graduates from this program staffed planning agencies in such major cities as Atlanta, Charlotte, Knoxville, and Columbia, as well as the local planning assistance programs in nearly all Southern states. Planning education programs at the University of Tennessee and in other Valley states were later established with TVA's help and staff cooperation.

*Public administration* The Southern Regional Training Program in Public Administration was a joint effort of the universities of Alabama, Kentucky, and Tennessee and TVA. Students spent part of their graduate program in each institution. Graduates from this program staffed many key positions in state government throughout the South. The program thus became a major force in improving administration of public programs in this region.

## Other TVA Regional Development Activities

The above are some but not all of TVA's efforts to create new institutions and to change programs of existing institutions so as to meet the challenges of the transition of the region into the national economy. Other programs dealt with such matters as public health, including the unique methods used to solve the malaria problem on the Tennessee River by raising and lowering water levels.

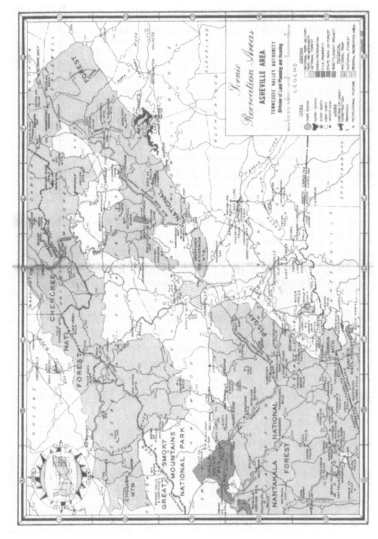

**Figure 2.18  Scenic Recreation Area, Asheville Area, Tennessee Valley Authority, Division of Land Planning and Housing, 1938**

*Source:* Tennessee Valley Authority.

Gordon Clapp, testifying before the House Appropriations Committee at the 1953 budget hearings noted that when TVA started 30 percent of the people in Alabama communities had malaria and after the malaria control programs of TVA and the States and localities were in place the incidence of malaria in 1948 had dropped to one-tenth of 1 percent. This technique contributed to solving the general malaria problem in the South. Other activities included water pollution surveys to pinpoint pollution sources; job training programs for employees on TVA projects; and regional library programs to help both TVA employees and communities near TVA projects.

## The End of an Era

The policies and programs which the first TVA Board adopted served TVA and the region well for the first twenty years. This was a period of remarkable achievement for TVA. In these years it had completed all the main river dams proposed in the 1936 report *The Unified Development of the Tennessee River System*. These structures provided a nine-foot navigation channel from Paducah on the Ohio river 650 miles up the Tennessee River to Knoxville, thus completing one of the specific mandates in the TVA Act. In addition, thirteen tributary dams were completed, which provided a major part of the flood control called for in the act.

TVA's remarkably rapid completion of its river improvement plan was spurred by the needs of the nation for electricity as it moved out of the Depression. Subsequently, there was a growing demand for power as a result of the defense needs of World War II. The system of dams and reservoirs also had to meet the growing demands from municipal, cooperative and large industrial customers within the Tennessee Valley Area.

In this twenty-year period TVA policies on regional development had contributed to the establishment of new or modified state and local institutions that were to help the Tennessee Valley and the Tennessee Valley states make the adjustments necessary for a successful transition from an agricultural and resource-based economy to an urban-industrial economy that more closely approached national patterns. These changes were reflected in the growth of manufacturing and other non-farm employment; in growing cities; and in altered patterns of agricultural and forest management and production. Tables 2.2, 2.3, and 2.4 show the dramatic changes in the in population distribution which occurred in the Tennessee Valley Region and the Seven Valley States over the period from 1930 to 1990. The trend was clearly discernible as early as 1950 and when compared to trends in the United States clearly indicated that the Valley and the Seven Valley States were starting to reflect population patterns similar to the United States as a whole. These trends should have indicated a need for TVA to reevaluate its regional development program to determine if it was meeting the needs of a growing urban-industrial economy. A look forward at these population trends shows that these changes would accelerate in the 1960s, 70s, and 80s when the population patterns of the Tennessee Valley Region and the Southeast would mirror the nation as a whole, TVA helped create and empower a wide variety of state and local institutions; among the most important were the State planning agencies. In 1935 expenditures for state planning activities in the Seven Valley States totaled $5,000, all in Alabama. By the eve of World War II the total had risen to $177,000 and all seven states had working agencies. And by 1956, the seven states had planning budgets totaling $1.3 million.

Had TVA planning and development activities succeeded? Not if the measure is some abstract vision of a physical and social utopia created from a detailed blueprint. But if the measure is the degree to which positive change and improvement in the condition of the people of the Valley occurred, TVA's first twenty years were a grand achievement, hardly equaled among other programs of the New Deal. The great physical artifacts that are the dams are impressive, but they should not obscure TVA's achievements in other areas – state and local government capacity building, industrial development, and parks. TVA also brought a skilled, talented and civically committed work force to the region, people who were used to involvement in local politics on a constructive level. The very presence of these educated outlanders, sometimes unwelcome among old-timers, nevertheless stimulated changes in the system of local government in many parts of the Valley.[6]

TVA's approach to regional planning and development was pragmatic not dogmatic; TVA defined its region not rigidly, but according to function. In effect, TVA planned within a multiplicity of regions: the river itself, the river drainage basin, the power service area, and the states in the Tennessee River Basin, the entire Southern Region of the US, and, for some purposes, the nation as a whole. The TVA staff worked within political realities, recognizing that planning and development, like politics, is the art of the possible. Of course, TVA could have been more effective in its planning and development and might well have achieved more had the enabling legislation been clearer, and had the ideas of the visionary Arthur Morgan continued to guide the Authority. But history and events unfold in ways not always within the power of officials or planners to shape or change. Nevertheless, TVA's original general plan was clear in the minds of its regional resource development staff and this plan directed their thought and actions. Much was achieved within that framework.[7]

The changes stimulated by TVA, coupled with the completion of the original plans for water improvement in the Tennessee River Basin, changed the very environment within which TVA operated. The regional improvement program had been closely tied to the river improvement program. With the completion of the river program, the future of TVA's regional activities came under new scrutiny, raising questions as to their future. The glory days for TVA planning seemed to be ending as the Authority started its second twenty years, searching for new directions better suited to the postwar period.

**Table 2.2  Tennessee Valley Region (201 Counties): Urban, Rural Non-farm, and Farm Population, 1930–1990**

| Population | Urban | Rural Non-farm | Farm | Total |
|---|---|---|---|---|
| 1930 | 1,253,474 | 989,713 | 2,721,879 | 4,965,066 |
| 1940 | 1,454,794 | 1,201,012 | 2,846,252 | 5,502,058 |
| 1950* | 2,041,124 | 1,608,920 | 2,262142 | 5,912,186 |
| 1960 | 2,670,507 | 2,264,426 | 1,246,591 | 6,181,524 |
| 1970 | 3,325,216 | 2,769,517 | 636,090 | 6,730,823 |
| 1980 | 3,968,287 | 3,536,971 | 337,008 | 7,842,266 |
| 1990 | 4,244,725 | 3,810,362 | 219,503 | 8,274,590 |
| | | | | |
| Change 1930–50 | +787,650 | +619,207 | -459,737 | +947,120 |
| 1950–70 | +1,1284,092 | +1,160,597 | -1,626,052 | +818,637 |
| 1970–90 | +919,509 | +1,040,845 | -416,587 | +1,543,767 |
| | | | | |
| Percentage Distribution | | | | |
| 1930 | 25.2% | 19.9% | 54.8% | 100.0% |
| 1940 | 26.4 | 21.8 | 51.7 | 100.0 |
| 1950* | 34.5 | 27.2 | 38.3 | 100.0 |
| 1960 | 43.2 | 36.6 | 20.2 | 100.0 |
| 1970 | 49.4 | 41.1 | 9.5 | 100.0 |
| 1980 | 50.6 | 45.1 | 4.3 | 100.0 |
| 1990 | 51.3 | 46.1 | 2.6 | 100.0 |
| | | | | |
| Change 1930–50 | +62.8 | +62.6 | -16.9 | +19.1 |
| 1950–70 | +62.9 | +72.2 | -71.9 | +13.9 |
| 1970–90 | +28.0 | +34.7 | -65.5 | +22.9 |

*New Urban Definition 1950–1990

*Source*: US Census of Population.

**Table 2.3   Seven Valley States\*: Urban, Rural Non-farm, and Farm Population, 1930–1990**

| Population | Urban | Rural Non-farm | Farm | Total |
|---|---|---|---|---|
| 1930 | 5,269,563 | 4,074,164 | 9,044,120 | 18,387,847 |
| 1940 | 6,158,013 | 4,719,113 | 9,274,218 | 20,151,344 |
| 1950\*\* | 7,788,843 | 7,142,747 | 7,370,817 | 22,302,407 |
| 1960 | 12,020,908 | 8,789,645 | 3,693,094 | 24,503,647 |
| 1970 | 14,975,389 | 10,340,805 | 1,807,497 | 27,123,691 |
| 1980 | 17,186,546 | 12,403,996 | 1,087,113 | 31,429,807 |
| 1990 | 20,253,012 | 13,538,581 | 678,902 | 34,470,495 |
| | | | | |
| Change 1930–50 | +2,519,280 | +3,068,583 | -1,673,303 | +3,914,560 |
| 1950–70 | +7,186,546 | +3,198,058 | -5,563,320 | +4,821,284 |
| 1970–90 | +5,277,623 | +3,197.776 | -1,128,595 | +7,346,804 |
| | | | | |
| Percentage Distribution | | | | |
| 1930 | 28.7% | 22.2% | 49.3 | 100.0% |
| 1940 | 30.6 | 23.4 | 46.0 | 100.0 |
| 1950\*\* | 34.9 | 32.0 | 33.1 | 100.0 |
| 1960 | 49.1 | 35.9 | 15.0 | 100.0 |
| 1970 | 55.2 | 38.1 | 6.7 | 100.0 |
| 1980 | 57.0 | 39.5 | 3.5 | 100.0 |
| 1990 | 58.7 | 39.3 | 2.0 | 100.0 |
| | | | | |
| Change 1930–50 | +48.0% | +75.0% | -18.5% | +21.3% |
| 1950–70 | +92.3 | +44.8 | -75.5 | +21.6 |
| 1970–90 | +35.2 | +30.9 | -62.4 | +27.0 |

\*\*New Urban Definition 1950-1990

\*Alabama, Georgia, Kentucky, Mississippi, North Carolina, Tennessee, and Virginia

*Source*: US Census of Population.

**Table 2.4  United States: Urban, Rural Non-farm, and Farm Population, 1930–1990**

| Population | Urban | Rural Non-farm | Farm | Total |
|---|---|---|---|---|
| 1930 | 68,954,823 | 23,662,710 | 30,157,513 | 122,775,046 |
| 1940 | 74,423,702 | 27,029,385 | 30,216,188 | 131,669,275 |
| 1950* | 88,927,464 | 38,693,358 | 23,076,539 | 150,697,361 |
| 1960 | 125,283,783 | 40,596,990 | 13,444,898 | 179,325,671 |
| 1970 | 149,334,020 | 45,586,707 | 8,292,150 | 203,212,877 |
| 1980 | 167,054,638 | 53,873,264 | 5,617,903 | 226,545,805 |
| 1990 | 187,051,543 | 57,786,747 | 3,871,583 | 248,709,873 |
| | | | | |
| Change 1930–50 | +19,972,641 | +15,030,548 | -7,080,974 | +27,822,315 |
| 1950–70 | +60,406,556 | +6,893,349 | -14,784,389 | +52,515,516 |
| 1970–90 | +37,717,523 | +12,200,042 | -4,420,567 | +45,496,996 |
| | | | | |
| Percentage Distribution | | | | |
| 1930 | 56.2% | 19.3% | 24.5% | 100.0% |
| 1940 | 56.5 | 22.0 | 21.5 | 100.0 |
| 1950* | 59.0 | 25.7 | 15.3 | 100.0 |
| 1960 | 69.9 | 22.6 | 7.5 | 100.0 |
| 1970 | 73.5 | 22.4 | 4.1 | 100.0 |
| 1980 | 73.7 | 23.8 | 2.5 | 100.0 |
| 1990 | 75.2 | 23.2 | 1.6 | 100.0 |
| | | | | |
| Change 1930–50 | +29.0% | +63.5% | -23.5% | +22.7% |
| 1950–70 | +67.9 | +17.8 | -64.1 | +34.9 |
| 1970–90 | +25.3 | +26.7 | -53.3 | +22.4 |

*New Urban Definition 1950-1990

*Source*: US Census of Population.

The experience of the first twenty years had provided, and continues to provide, important lessons for regional planners everywhere, particularly in developing countries facing problems similar to those that confronted the Tennessee Valley.

## Notes

1　One important exception is the Port Authority of New York and New Jersey whose origins and history somewhat parallel those of the TVA. For a comprehensive overview of this agency, which still plays an important role in the New York Metropolitan Region, see Doig, J W. (2001), *Empire on the Hudson: Entrepreneurial Vision and Political Power at the Port of New York Authority*, Columbia University Press, New York

2　See Appendix for correspondence relating to the formulation of the Regional Planning section of the original TVA Act

3　For a critical view of Morgan's philosophy see Ray Albert, Jr., FDR's Utopian, Arthur Morgan of TVA, 1987, (Jackson, Mississippi and London: University of Mississippi Press). For a sympathetic view, see Ernest Morgan, *Arthur Morgan Remembered*, 1991, (Yellow Springs, Ohio: Community Service, Inc.)

4　Draper, Earle S. 1983. (Address presented at the City of Norris Fiftieth Anniversary Celebration, October 14, 1983.) Personal sound archives of David A. Johnson

5　For a perceptive insight on this problem, see the commentary by Guthrie S. Birkhead of the Maxwell School, Syracuse University, in Government in Metropolitan Areas: Commentaries on a Report by the Advisory Commission on Intergovernmental Relations, 87th Congress, First Session, December 1961 (Washington, U. S. Government Printing Office, 1962): 87–93

6　For a more detailed account of TVA's contribution to Southern planning and development, see Albert Lepawsky, *State Planning and Economic Development in the South* (Kingsport, Tennessee: National Planning Association Committee of the South, Kingsport Press, August 1949): 132–139

7　For more detailed information on TVA's regional programs, see Florida State Lectures, 1953, in TVA Technical Library; this is a series of lectures given by key TVA staff. Also see Gordon R. Clapp, *The TVA: An Approach to the Development of a Region* (Chicago: University of Chicago Press, 1955); this is a published version of the Walgreen Lectures given by Mr. Clapp at the University of Chicago in February 1954. For a list of the Memoranda of Agreement see TVA Cooperative Relationships, 1946, 86 pages, mimeograph in TVA Technical Library. This last reference is an interesting document because it shows the great diversity of cooperative relations maintained by TVA

# Chapter 3

# TVA's Second Twenty Years: The Search for New Directions in the Regional Planning and Development Program, 1953–1973

## Introduction

TVA's third decade, beginning in 1953, started on a high note. TVA had major achievements to its credit; both the river improvement and power programs had exceeded expectations and the overall regional development program had been reasonably successful and had widespread acceptance in the Valley and the seven Valley States. Moreover the economy of the region was starting to reflect national conditions. But these achievements also brought on new challenges for TVA, some growing out of the success of its programs and others resulting from changes in the conditions and needs of a growing region.

During the next twenty years the TVA regional development program was to be influenced by three factors: the growing dominance of the power program in the overall TVA program; the completion of the major elements in the river improvement system, and the changes in the demographic and economic characteristics of the Tennessee Valley and the entire South.

## The Growing Dominance of the Power Program in the Overall TVA Program

During its first twenty years TVA had managed to meet the seemingly insatiable power demands of farms, residences, industries, and national defense activity in the region. At first it was national defense which was the driving force for a larger and more reliable electric generating capacity. It all started almost unnoticed in 1940 when it was determined that 'the TVA system as planned is inadequate to meet the requirements established by the Advisory Commission of the Council of National Defense.' Additional generating capacity was needed in the eastern part of the Valley to support the existing hydro facilities. The site chosen for the first steam generating station to be built by TVA was near Watts Bar Dam.[1] By present day standards the Watts Bar Plant was small; four 60,000 kw units for a total of 240,000 kw.

The first unit was started in August 1940 and the last unit placed in operation in April 1945.

World War II resulted in increases in the capacity of the TVA power system. The first unit of the Watts bar plant came on line in 1942. Cherokee Dam was built in sixteen months and Douglas Dam was built in a year. These two projects added 235,000 kilowatts to the TVA system. The availability of large amounts of economical power was a major factor in the location of the Oak Ridge atomic plants. In addition, the TVA fertilizers plants at Muscle Shoals produced 50 percent of the phosphorus used by the Armed Forces during World War II. The demands for power continued to grow after the war. TVA responded supported by both the President and the Congress. Johnsonville Steam Plant was started in 1949; Widows Creek in 1950; Colbert and Kingston in 1951; John Sevier in 1952 and Gallatin in 1953. In 1953 'TVA had under construction . . . 50 generating units at 17 dams and steam plants. These facilities added 1,145,600 kw of generating capacity to the system, by far the largest addition to the generating capacity to be put in operation by TVA in a single year.' There were also additions of 90,000 kw from the U.S. Corps of Engineer Projects on the Cumberland River. The result was that the installed capacity increased to 5,102,985 kw by June 30, 1953.[2] In fact, by 1954 the system capacity had doubled in two years. With an additional 4,856,000 kw under construction the anticipated capacity by December 1956 was to be 9,9938,985 kw[3] (Figure 3.1).

To illustrate the changes which had taken place in the TVA system, in 1939 almost all TVA power was produced at hydro stations; but by 1956 seventy percent of TVA power was generated at steam plants. Recalling the concerns of Congress and the speculations of the private power industry that there was not a market even for the generating capacity planned into the hydro system, the growth of the TVA power really reflected the economic changes which had taken place in the region. But the rapid expansion also brought new problems to TVA in the form of growing pressure from the private electric industry and the Congress to take a new look at how this huge and growing power system would be financed. The change of the national administration and the strength of the TVA support in Congress was to play a large part in the outcome.

The struggle over financing the TVA power facilities started in 1948 when TVA requested funds for a large steam generating plant to be built at New Johnsonville, Tennessee. Senator Albert Gore, Sr., a staunch TVA supporter, proposed an amendment to the TVA appropriation act as recommended by the House Appropriations Committee to include funds for the New Johnsonville Steam Plant. The budget proposed by President Truman had included funds for the project. The House with 245 Republicans and 185 Democrats voted 192–152 to reject the amendment. The Senate with 51 Republicans and 45 Democrats voted 45–37 in favor of the amendment but in conference the House refused to discuss the amendment and thus it was defeated. However in the surprise election of 1948 not only was President Truman reelected

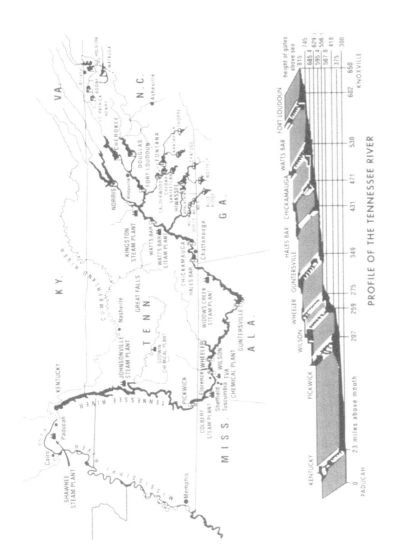

**Figure 3.1   TVA Dams and Steam Plants on the Tennessee River, 1950**

*Source:* Personal files of A.J. Gray.

but the Democrats gained control of both houses of Congress; in the House (263–171) and in the Senate (54–42). Funds for the New Johnsonville Plant were approved in 1949.

The power demands of the national defense build-up resulted in the huge appropriations for six giant steam plants. By 1952, however, the congressional support for additional funding of TVA steam power stations narrowed and votes in the Congress seemed to show that congressional members from southern states outside the Tennessee Valley tended to refuse funding for such plants. Eisenhower took office on January 20, 1953 and his administration immediately started a review of TVA requests for more appropriations for steam plant construction. There were growing demands for self-financing of such construction.

In an extensive interview in August 1953 with *Knoxville News-Sentinel* reporter Julian Grainger, TVA Board Chairman Gordon Clapp stated the official TVA position by rejecting any 'suggestions that TVA float revenue bonds to finance future expansion of its electric power system'. Clapp maintained that 'such a scheme . . . would merely give TVA's foes a "wedge" that could force "mismanagement" and perhaps destroy the Authority's "ability to maintain low-cost power production"'.[4] He did not realize how true his prophesy would be for TVA in the 1970 and 80s.

TVA was now faced with a dilemma. With diminishing congressional support, a President who supported a partnership principle rather than continued TVA appropriations for steam plant construction, and TVA's own policy decisions to support only appropriations for steam plant expansions, there was still a gap between the capacity of the TVA system and the power demands from both the defense programs and domestic needs. TVA chose to request funds to build a plant at Fulton, Tennessee, a few miles north of Memphis. Although Truman rejected the recommendation in his 1952 budget, it was included in his last budget message of January 9, 1953.

The incoming Eisenhower Administration responded with the Dixon-Yates proposal. Under this proposal Middle South Utilities, Inc. headed by Edgar H. Dixon and the Southern Company headed by Eugene A. Yates agreed to form a separate Mississippi Valley Generating Company (MVGC). Under contract with the Atomic Energy Commission (AEC) MVGC would replace the 600,000 kilowatts being provided by TVA by turning over the same amount of kilowatts to TVA. This proposal would have given a private utility a generating station within the TVA service area and would provide power to a major TVA customer thus freeing up power for TVA to use for domestic needs of the power service area. The proposal had strong support within the Eisenhower Administration, the Congress and, of course, the private power industry which saw this a first step toward selling the TVA power operation to private utilities. Two events were to influence the outcome of this public debate. First there were serious allegations of conflict of interest on the part of both the government agencies and the private companies involved. The

**Figure 3.2   Gordon R. Clapp, headed TVA Personnel Division; General Manager, 1939–1946, TVA Board Chairman 1946–1954. Made major contributions to the TVA regional development program by articulating guiding operational principles for the staff**

*Source*:   Patricia Bernard Ezzell, TVA Historian.

result was that the Eisenhower Administration withdrew its support from the proposal and the Dixon-Yates contract was abandoned.[5]

Second, a change in the TVA Board brought about significant change in TVA policy which affected the financing of the TVA power program. Gordon Clapp's term as a member of the TVA Board expired May 18, 1954 and Army General Herbert D. Vogel was appointed Chairman of the Board. He took office September 2, 1954. Although he had supported the Dixon-Yates proposal, he also favored TVA's constructing its own generating facilities with TVA revenue bond financing to pay for such facilities. The amendment to the TVA Act to permit bond financing of power facilities was signed into law on August 6, 1959 and the first power bonds were issued on November 15, 1960. The amendment also limited the TVA power service area to that which existed July 1, 1957. As TVA was to discover later, the amendment did not prevent private power companies from expanding into the TVA service area.

This change in the TVA power financing policy was to change the balance between power and non-power programs and thus the nature of TVA's regional development program. Power became the 'paying partner' and the major consideration in TVA policy matters. As Roscoe Martin noted in 1956 'the production and sale of power have come to dominate program action if not program thinking in TVA'.[6] In 1957, for example, out of a total budget of $186.6 million, $177.6 million was allocated to the power program with but $9 million going to all non-power functions. In fact, of the $1.9 billion which had been appropriated to TVA up to 1956, $1.3 billion was spent on power.

This should not be too surprising considering that most of these moneys were spent on physical facilities such as dams and reservoirs, generating stations, transmission lines and substations. Although such activities took most of the TVA funds, to compare moneys spent on facility construction with other aspects of the regional programs which involved primarily technical staffs is not a comparison which recognized the differences in the activities. Only a small number of professionals from agriculture, forestry, community development and planning, industrial development, public administration and finance, were involved in the regional development program as compared to those involved in construction and operation. Although national defense initially may have been a major force in the growth of energy production, the changing economic and population distribution patterns in the Tennessee Valley Region as well as the Southeast as a whole was the sustaining force in power demand. It was in fact the result, at least in part, of TVA's fostering of industrialization not only in the Tennessee Valley but also the entire South.[7]

## Completion of the Major Elements in the River Improvement System

It is important to recall the role of power in the regional development program as proposed by the original TVA Board. The availability of inexpensive and abundant power was one of the forces moving the Tennessee Valley and the South toward the kind of urban industrial society that characterized the rest of the nation. But the TVA power program also supported the overall regional development program because it helped establish the concept that power was a part of the larger TVA regional program rather than TVA being a single purpose electric public utility.

By the mid 1950s, TVA had virtually completed the water improvement program as envisioned in its 1936 report to Congress, *The Unified Development of The Tennessee River System*. In fact, 1954 marked the first time in 21 years that TVA did not have a major system dam under construction. Under policies established by the original TVA Board, the many facets of the regional program were related, although in some cases tenuously, to the river improvement program. This relationship was particularly important in requests for appropriations to fund these programs.

## The Changing Demographic and Economic Characteristics of the Tennessee Valley and the South

The demographic and economic patterns which had characterized the South since the nation was founded were also changing. Within the 20 year period between 1930 and 1950 the non-farm population in the Tennessee Valley Region grew by 1.4 million while the farm population declined by nearly one-half million. This paralleled the trends in the seven valley states. As might be expected the non-farm population in both regions was growing at a faster rate than in the United States as a whole (See Tables 2.2, 2.3, and 2.4).

That the Valley and the South were becoming urbanized and that the future of both regions rested on its cities and industrial development was clear by 1960, perhaps even earlier. For example, between 1930 and 1970 the urban population of the Tennessee Valley Region grew by over 2 million and the rural non-farm population, located mostly in the suburban fringe grew by nearly 1.8 million. These changes were largely a response to the change in the ways the people of the Valley earned a livelihood. In 1929 fifty-five percent of the workers in the Tennessee Valley Region were employed on farms. By 1970 that percentage had been reduced to 9.5 percent while over 90 percent of the workers were employed in non-farm jobs.

Another comparison shows how rapidly the Valley was changing in ways that reflected employment patterns similar to those of the nation. In 1929 when 55 percent of the Valley's workers were employed on farms less than 25 percent were so employed in the nation as a whole. By 1970 the percentages were 9.5 percent for the Valley and 4.1 percent for the nation. It was these changes in the social and economic life of the region along with the virtual completion of the river improvement program and the growing dominance of the power program within TVA that were to shape the effectiveness of TVA's regional development program in its second twenty years. Within this context

TVA began its search for a new and unifying principle to reshape its regional development program.

## The Search for New Directions

In its first twenty years TVA had influenced significantly improvements in both the base and productivity of the agricultural, forest and other resources of the Tennessee Valley and the Tennessee Valley States. Although little noticed, TVA also made important contributions to the improvement in the quality of urban development in the South and, as will be noted later, in the nation as a whole. Both natural resource management and urban improvements were a direct result of TVA efforts to develop state and local institutions capable of dealing with changing resource and urban problems. Institutional development was a major and unheralded contribution of TVA to the Tennessee Valley Region and the Southeast.

But TVA's own reassessment of its place in the region did not give adequate attention to the fact that by 1950 the Tennessee Valley Region as well as the South was changing to an urban industrial society and was even then tending to mirror national trends. Instead TVA continued to place its major program emphasis on rural resource development and relatively less attention to programs which would have influenced urban growth and development.

## The Tributary Area Development Program (TAD)

A little known program in a small watershed development was to have a major influence on the search for new directions for TVA's overall regional resource program. Recalling again that the non-construction and non-power parts of the regional resource development program depended heavily on the river improvement program as a justification for appropriation requests, it is not surprising that TVA looked at this watershed program as it searched for a new and hopefully continuing basis for these requests.

In 1951 work started in the 134 square mile Chestuee Watershed located within the Great Valley of East Tennessee and about midway between Knoxville and Chattanooga. There was general consensus within the TVA natural resource staffs and to some extent within the Board itself that if the Valley-wide program to encourage soil, water, and forest conservation and development on a broad scale could reverse the downward spiral of resource depletion, then the same kind of concentration would produce results in small but problem-ridden watersheds. In the Chestuee Watershed detailed surveys were made of farms and other small tracts by agriculturists from the University of Tennessee Extension Service and the State Experiment Station and by TVA foresters. These surveys provided information on land resources and land use practices which became the basis for planning for needed land use changes. An advisory committee was established in the watershed to work with the state agencies and TVA.[8]

The lead staff for TVA efforts at watershed development was the Division of Forestry Relations. As early as 1949 the forestry group was working on the reforestation of several badly eroded small watersheds in west Tennessee and northeast Mississippi. By 1951 TVA had designated the 300 square mile Beech River watershed in Decatur and Henderson counties in west Tennessee as the second watershed in which to make comprehensive studies of resource use and development. By 1953 reconnaissance surveys were made in small watersheds in other Tennessee Valley states.[9]

At the same time TVA was working on a general concept to use small watersheds as a basis for dealing with areas having resource problems and conditions which seemed to require special attention. In its 1953 report *Working with Areas of Special Need with Examples from the Beech River Watershed* TVA rationalized the watershed approach in these words:

> Among these areas of special need or opportunity, tributary areas are in many aspects unique. Because their problems center around land-water relationships the solutions may seem to lie in land-use changes and water control. Actually, they often present in capsule form something of the many-sided problems as a major river basin. The approach to tributary watersheds, to be successful, cannot be centered in forestry development plans alone, nor in agriculture, nor in industrial development. Their needs are complex and their opportunities diffused.[10]

In retrospect this move to emphasize the watershed as a basic planning unit is difficult to understand particularly when the strength of the earlier regional resource program seemed to rest on TVA relationships with the states and the larger seven state region. In the first twenty years TVA had recognized that the river basin was the appropriate region for planning water resources but that larger regions were necessary to deal with other matters of broad regional concern.

The attempt to use the watershed as an overall planning unit certainly was counter to the then current trends in both regional planning thought and practice. TVA actions tended to support Melvin R. Levin's comment on TVA that 'the danger with the 'river basin' approach is its unsuitability for an urbanized nation in the present era. It seems questionable that a region created on the basis of natural resource needs and problems can focus adequately on urban problems'.[11]

In its 1961 Annual Report TVA also seemed to recognize the problem raised by Levin when it noted:

> The region's weakest points are its overpopulated rural areas where thousands of farms are still too small for economic operation or adequate incomes. These areas are the source of rapid migration from farm to city and from the Valley to other sections of the United States where employment opportunities seem more numerous. Despite steady progress in improving agricultural resources, measures

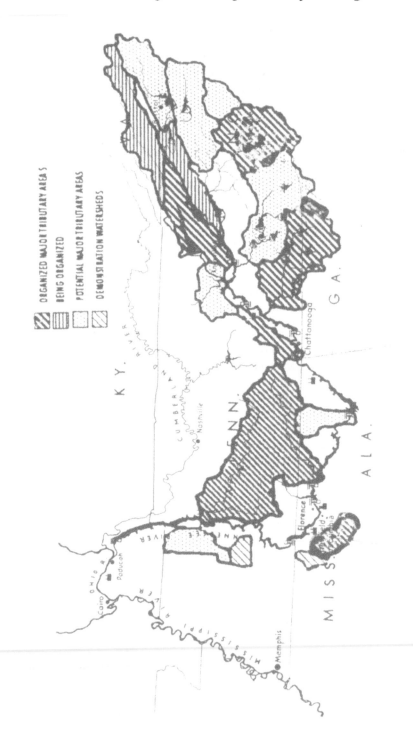

**Figure 3.3  Tributary Area Development in the Tennessee Valley, Present and Potential Operations, TVA, June 1962**

*Source:*   Tennessee Valley Authority

are needed to speed the increase in the size of farms and making farming more profitable. Despite good progress in industrialization and commercialization, steps are still needed to absorb more rapidly and more completely the people who must shift from rural to urban occupations.[12]

The report then stated that a Tributary Area Development Office had been established and that this organization would provide a new administrative vehicle for the collaboration of TVA divisions in order to facilitate the shift from rural to urban occupations. Although this seemed to recognize the growing urbanization of the Valley it should be noted, however, that TVA still worked within the confines of a small watershed and that the TAD office was headed by a trained forester. With the organization of the Office of Tributary Area Development, TAD began to encourage the formation of watershed associations throughout the Tennessee Valley. Figure 3.3 shows the areas organized as of 1962.

At first TVA insisted that the tributary watersheds and the citizen associations organized to plan watershed programs would be concerned with overall watershed development, not water resources alone. The tributary development associations supported the TVA premise as to the broad purposes of the TAD program. But some of TVA's friends were becoming concerned with the lack of action and new economic development in the tributary areas. There were growing demands that TVA build dams in these basins. In fact, both the *Nashville Tennessean* and the Louisville *Courier-Journal*, long time friends of TVA, criticized TVA for not proceeding with a major program of dam building in tributary areas.[13] In fact, TVA started building such structures in 1962.

As it turned out, the building of dams in the tributary areas was soon to became a secondary issue in the program. In the early 1960s President Johnson's 'Great Society' programs were in full ascendancy. Rapidly expanding Federal grant programs affected most aspects of state and local government including water supply, transportation, housing, waste disposal and sewerage, recreation and open space, economic development, and health, to mention but a few. The traditional method of channeling federal money through the states to local governments was based on the vertical functional cooperation that existed between federal, state, and local agencies. With the beginning of the era of the large federal grants directly to local agencies this tradition was weakened.

The Department of Housing and Urban Development (HUD) was one of the first of the major federal departments to seek new ways to deliver its programs. HUD believed that there had not been an adequate response to urban interests by what HUD considered were rural-oriented state legislatures and state executive departments. With the huge sums of money involved federal agencies already dominated the functional areas for which they were responsible. But to add to that power many agencies wanted to have a 'grass roots' constituency of their own. In addition each Federal agency had its own program regulations and many came to require a regional organization as a condition for receiving grants. Suddenly multi-county regions were popular.

The result was a confusing system of overlapping special purpose regions which soon became a major issue in the Congress. Finally the Bureau of the Budget (now the Office of Management and Budget) issued Circular A-80 which called upon the governors to set up for each state a single system of sub-state regions and it directed that all federal agencies use such regions.

But the TVA Office of Tributary Area Development was faced with the problem of uniform sub-state regions even before these mandated federal regulations took effect. In 1965 the act creating the Appalachian Regional Commission (ARC) was passed and its regulations required states to establish a system of local development districts within the ARC region. The Economic Development Administration (EDA) was created the following year and it required economic development districts as a condition for grants. However, the Tennessee Development District Act of 1965 preceded most of these actions by the federal government. This Act required the Tennessee State Planning Commission to work with localities in establishing a single system of multi-county regions within the state and was adopted because of the difficulties the state and local governments were having in trying to respond to the federal demands for separate functional districts or regions. This Act also gave the state a program to encourage the individual Federal agencies to use the state sponsored multi-county districts. Too, Tennessee was a strong supporter of ARC and had anticipated its regulations requiring substate districts. Since TVA had already encouraged the establishment of several tributary area development associations this action by the state of Tennessee raised questions about how the TVA tributary area development program would relate to the sub-state districts established under the new legislation.

Within TVA there were differing staff views as to the position TVA should take regarding the Tennessee sub-state development districts. The Regional Studies Staff had long supported active state planning organizations and was the principal advocate for an urban approach to valley development problems. In a February 24, 1966 memorandum to the TVA General Manager the Director of Regional Studies noted that 'Not only must TVA contend with the regional expression of the new programs set up by functional agencies in Washington, but it must also be cognizant of the state planning efforts aimed at the delineation of state economic areas, the criteria for which may not coincide with TVA's own subregional programs. Although TVA's TAD program preceded the newer regional activities originating in Washington or in state capitals by many years, it is no longer possible, in my estimation, to ignore the problems of overlapping areas and jurisdictions'. The memorandum continued by pointing to the growth of manufacturing jobs and the expanding urban centers and concluded 'Thus, economic and social associations are more easily delineated on a trade area or commuting area basis than around small watersheds where community interest is primarily related to natural resource-use issues'.

The Regional Studies Staff also prepared a report entitled 'TVA and the Emerging Economic Development Districts' (April 21, 1966) which urged TVA to support the sub-state districts and to make a concerted effort to rethink the objectives and performance of the TVA sponsored watershed agencies as well as relation of the watershed agencies to general state and local governments.

TVA, however, decided to hold to the Tributary Area Development Associations, and particularly in Tennessee, tried to merge the TAD areas into development districts. This was to start a general deterioration of TVA-State relations which had been the centerpiece of the TVA regional development program. For example, in inviting TVA to a meeting to discuss the formation of an economic development district for East Tennessee the Tennessee State Planning Agency stated 'All Federal agencies will be expected to cooperate and assist in making the district a success'.

The TAD program was also having problems in other states. Following devastating floods in August 1940 TVA began to prepare a flood control plan for the North Carolina portion of the French Broad Watershed. World War II intervened, but in 1950 TVA submitted a report to the President entitled *A Flood Control Plan for the French Broad Valley.* The plan included detention dams, levees, and channel improvements and carried a price tag of $21.5 million. Agricultural benefits constituted the bulk of the benefits. By 1956 the cost had risen to $27.5 million.

Two problems complicated the decisions on how TVA might proceed with this project. First, the USDA's Soil Conservation Service had been active in the area and had proposed a plan for one of the French Broad tributaries that would give protection to land proposed for a detention basin under the TVA plan. Second, although there was some support in the region for the TVA plan, it was a qualified support which indicated local desires to have the plan modified. TVA recognized that Soil Conservation Service activity in the French Broad Basin would 'increase the difficulty in obtaining area-wide acceptance of the TVA Plan'. In addition, because of the lack of area-wide support TVA seemed reluctant to spend the additional moneys needed to revise the plan so as to meet area needs as viewed by local communities.[14]

In the late 1950s the Western North Carolina Regional Planning Commission had been organized to include the 12 counties centering on Asheville. The TVA French Broad Valley Flood Control Plan affected only three of the counties within the jurisdiction of the commission: Buncombe, Henderson, and Transylvania. To try to resolve these problems TVA decided to review the TVA Flood Control Plan with the regional planning commission.[15]

TVA meetings with the regional planning commission clarified major short-comings of the TVA plan as viewed by local people and some of the reasons for the lack of area support for the plan. As noted above the TVA plan was based almost exclusively on agricultural benefits whereas the over-all local area interest also included municipal and industrial water supply, protection of industrial sites, and recreation. To respond to such broadened interests it was proposed that the regional planning commission set up a special planning group within the French Broad basin to look into the over-all water needs of the area.[16]

Although this approach was generally acceptable to TVA, major policy makers were still having difficulty in placing the flood control plan within a context of over-all area development. Staff was reminded that TVA had given much thought and spent considerable money on the TVA plan and that any study group should not start all over again as if a flood control plan did not exist.[17] In addition, TVA, particularly at the major policy level, seemed to be having difficulty adjusting to the fact that the Asheville Basin had changed considerably since the 1940 floods; the basin was much more urban and business and civic leaders saw its future in commerce, industry, and recreation rather than in agriculture.

The TAD Staff became involved and following the practice used in tributary areas encouraged the organization of a local citizen group to study and review the project. In March 1961 the Upper French Broad Economic Development Commission was established. The commission and TVA encouraged state agencies with interests in water and area development to review the project and provide the commission with their views. What the commission and TVA found were widely differing views and loyalties among these agencies.

The North Carolina Department of Water Resources was the principal state agency concerned with water resource development but because of its limited staff took the position it could not make a critical appraisal of the TVA plan and thus could not support or reject the plan. The State Recreation Board strongly supported the plan. The Department of Conservation and Development interests were expressed primarily through its divisions of Community Planning and of Commerce and Industry but neither took a strong position for or against the plan. Both the Wildlife Commission with its strong ties to sportsmen's groups and the State Soil and Water Conservation Commission with ties to the Soil Conservation Service opposed the plan.

The only state group capable of coordinating these varied state interests was the State Planning Task Force which had been set up within the Department of Administration. But it became involved too late in the process to be effective. Moreover there was clearly a need to broaden the planning base within the region itself so that a proposed development program would take into account not only the conventional water development concerns such as flood control, stream augmentation, and recreation but also the concerns of wildlife, urban, and industrial development interests. Such an approach would have helped to determine what changes might be made in the TVA flood control plan to respond to the interests of those concerned with urban and industrial development, recreation, the development of natural streams and the protection of fish and wildlife resources.[18] It was concern over protection of natural streams and fish and wildlife resources which produced the greatest opposition to the TVA flood control plan.

Ultimately, the State of North Carolina, the local agencies, and TVA were not able to find the common ground necessary to make such a broad plan acceptable to all the interests concerned and the TVA flood control plan for the Upper French Broad River was abandoned. It might have worked if the interests of those concerned had been involved in the preparation of the plan but at the time of the 1940 floods such a participatory process was not understood or appreciated.

Although the TVA Tributary Area Development Programs continued to be expanded into other parts of the Tennessee Valley through the 1960s opposition from environmental groups and changes in the Board resulted in the termination of the program in the mid-1970s. This effort by TVA to redefine its regional development mission failed primarily because it did not recognize the changes that had occurred in the Valley and the South and also perhaps because it was unwilling to look at the TVA Act and propose major changes in TVA responsibilities; an action which would have required opening the TVA Act to amendment, something TVA was always reluctant to do.

## Valley and Region-wide Resource Development Programs

The TAD program was a major departure in development policy for TVA. TAD started to turn TVA inward by encouraging TVA-sponsored citizen groups to handle development problems rather than working through the traditional state and local governmental units. This policy change caused strains in the relationships between TVA and state and local governments. It should be noted, however, that even though the major thrust of TVA's regional development program during the second twenty years was represented by TAD the regular developmental programs continued much as they had during the first twenty years although there were changes in resource problems.

### Agriculture

During TVA's second twenty years both the agricultural and forestry development programs continued to use improved technologies and education to make the land more productive. The National Fertilizer Center was established at Muscle Shoals, Alabama and it became a major source of information on new fertilizer technology in the country. Here new fertilizer products and manufacturing processes were perfected. The use of the products and processes were in turn demonstrated to farmers and to the fertilizer industry. The testing of the new fertilizers was carried out under actual farm conditions in 48 states and, perhaps more important, over 300 fertilizer companies were licensed to use the processes to manufacture and sell the products developed in TVA laboratories.

*Forestry*

Since at least the 1930s nearly sixty percent of the land area of the Tennessee Valley. has been forested. The earlier efforts of TVA and the states to control forest fires and to reforest marginal and eroded land were replaced in the 60s and 70s with programs that stressed forest management so as to improve production and to better utilize forest products. In addition TVA started experimental programs for the propagation of better and faster growing trees including methods for reproducing such improved forest stock.

*Wildlife*

The program to identify the TVA land suitable for wildlife programs and the transfer of such lands to appropriate federal and state agencies was completed during the 1970s. Over 86,000 acres of TVA land and water are now devoted to federal wildlife refuges on Kentucky and Wheeler reservoirs. Nearly 104,000 acres were made available for state wildlife programs in Alabama, Kentucky, Mississippi, and Tennessee.

*Recreation*

During its first twenty years TVA completed surveys of the recreation resources of the Valley, identified TVA lands along reservoirs suitable for parks, public access areas, and other recreation uses both public and private. It also worked to encourage state and local governments to create recreation agencies to develop and manage these resources.

During its second twenty years TVA emphasized development of these recreation resources. The result was that by 1972 TVA reservoirs shorelands included 17 state parks, 37 county parks, 42 municipal parks, and 423 public access and roadside parks of which the greatest number, 377, were under the control of state agencies. In 1953 alone TVA transferred to Tennessee 226 waterfront public access areas on nine reservoirs, the first time a state had assumed major responsibility for assuring the right of the public to have access to Federal reservoirs.

In 1969, however, the recreation staff recommended that TVA began to develop and manage recreation facilities at 43 day-use and camping areas on reservoirs and at dam reservations. These were located on 19 main river and tributary reservoirs and at one natural area near Tracy City, Tennessee. This change in policy was the result of TVA's dissatisfaction with the progress made by the states, particularly Tennessee, in making improvements and providing maintenance to public access areas. This, too was to disrupt and cause strained relations between TVA and state recreation agencies.

TVA also started a study that would lead to the creation of the Land Between the Lakes demonstration recreation area; a 170,000 acre area located between Kentucky and Barkley lakes. In June 1961 the results of the study were sent to President Kennedy with the recommendation that the area be made a national recreation area under the control of the National Park Service. The concept is shown in Figure 3.4 The park service concluded that the project did not fit into its scheme of national recreation areas. President Kennedy then turned the project over to TVA for development. TVA began construction of Land Between the Lakes in 1964. Located within 500 miles of 70 million people the area attracted over one million people in 1969. It remains a popular recreation and conservation education area serving the middle part of the country.

## An Urban Development Program – A Small Beginning

The difficulties with the TAD program were the result of TVA's failure to understand the nature of the changes taking place in the Valley and the important interrelationships of its own programs. As early as 1962 the Regional Studies Staff had made recommendations to the TVA Board and management that the changes which had occurred in both employment and population distribution patterns required more attention in TVA program planning:

> . . . the urbanizing areas of the Valley region now and in the foreseeable future are the major areas of employment and growth. The United States now has a 'national' economy. It is a unified whole, not merely the sum of widely dispersed 'big' and 'little' economies. The manufacturing and service industries which account for most of the growth are major factors making for urbanization. This characteristic of the economy brings into sharp focus the free competition on a national scale among urban-industrial areas for new industry. The result in the Valley is that a satisfactory adjustment of people to employment opportunities will require improvement of the competitive position of urban areas of the Valley relative to those outside the area. In addition, improvement of depressed areas is more a function of the development in nearby urban areas than of programs devised to develop small 'self-sustained' economies.[19]

The Staff also suggested that the time had come for TVA to reassess its role in the region and to prepare a new 'plan' for the Valley that would replace the 1936 report to Congress entitled *The Unified Development of the Tennessee River System*. There was considerable overall TVA staff support for this proposal but after much discussion it was rejected by the TVA Board.

The TVA rural resource utilization improvement programs which correctly stressed the new technologies with larger management units and increased

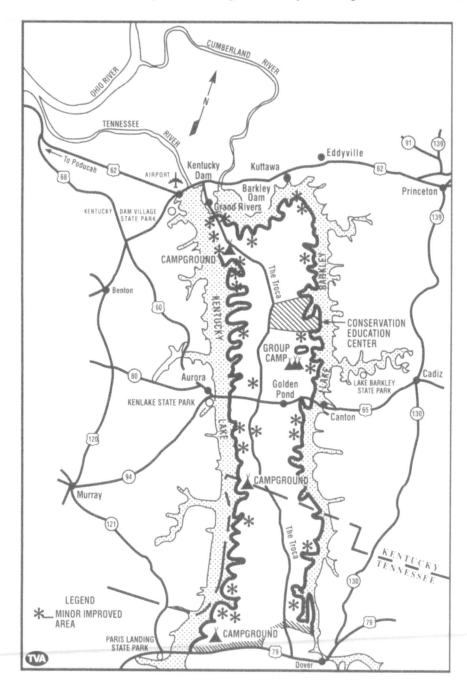

**Figure 3.4   Land Between the Lakes, map, 1964**

*Source*: A.J. Gray files.

incomes had the effect of forcing more and more people off the farms. As the tide of people moving off the farms grew larger and larger TVA's weak and understaffed urban and industrial development programs could not generate enough new industrial and urban service jobs to meet the demand. To put the problem another way, at a time when TVA needed a strong urban development policy that would emphasize an improved infrastructure, an integrated transportation system for the region, and new institutions with capabilities necessary to meet new urban and industrial challenges TVA was stressing industrial development in rural areas through its TAD program. This was not the best match to conditions that had emerged in the region.

At this point in time TVA also had an opportunity to mount what could have been a national urban development demonstration that would stress a regional system of cities that were sound economically and socially manageable. Such a demonstration would have been particularly important in the Tennessee Valley and the entire South because urban growth in these areas was coming at a time when the automobile was the principal form of transportation and was having a major impact on the form and structure of Southern cities.

Suburbs spread further and further from the large and small cities. The more widely dispersed population distribution patterns created problems for water and sewer services, schools, recreation, police and fire protection, and required new methods for service delivery.[20] Instead of responding to this new opportunity to help build for more efficient and satisfying urban living TVA continued to place its major program emphasis on resource and rural development which had served it so well during its first twenty years.

During the 1960s and 1970s TVA did continue to assist Valley states on matters of local planning assistance. In addition there were some special and significant new programs which were important to the development of the Valley region and in some cases the Nation as a whole. One example is the flood damage prevention program.

## Flood Damage Prevention

Perhaps the most significant new program during this period was the TVA work in flood damage prevention. This program really started in 1950 when the TVA Chief Engineer released a report entitled Major Flood Problems in the Tennessee River Basin.[21] This report noted that many communities upstream from reservoirs had serious flood problems but because there was not yet sufficient development in the flood prone areas to produce a favorable cost-benefit ratio flood control projects could not be justified. The only remedy proposed was that TVA establish a flood warning system throughout the Valley to warn people living in flood plains of impending disasters.

In reviewing this report and related internal TVA memoranda it is interesting to note that the report must have been sent to Gordon Clapp, a non-engineer, who at that time was Chairman of the TVA Board. Mr. Clapp's reaction to the report was 'What should TVA do: wait for development of the flood plains so that a flood control project could be justified?'. He recommended that the report be circulated to other TVA staff particularly to the Division of Regional Studies to get other staff reactions and possible alternative approaches to the problem. It is interesting that the response of the Chief Engineer cited the difficulty in obtaining community interest in the problem and that 'The subject of floods and flood control is so specialized and complex that it would be very difficult for someone not fully acquainted with these complexities to convey the proper picture to another, who is likewise unfamiliar with the subject.' He concluded with his belief that 'the matter is well in hand and our personnel are frequently in contact with public officials most interested in this matter'.[22]

The Regional Studies Staff after its review of the report proposed an entirely different approach to the problem. This staff started with the proposition that the problem in areas subject to flood related primarily to the control of land use and development and that the state and local planning agencies must have a key role in any regional or local effort to minimize flood hazards. The staff also noted that a review of planning literature and programs revealed that not much was being done on the application of stream flood and flood data to community plan development or to such land use controls as zoning and subdivision regulations. It did discover that Dr. Gilbert White of the University of Chicago Geography Department had proposed such applications and that the only real effort to deal with this problem was in the 1930s when an attempt was made to exclude all development from areas subject to flood. The staff did not consider this a realistic approach but to date no one had proposed an alternative. Finally the staff rejected the Chief Engineer's recommendations as inadequate to meet what it considered a growing flood problem in Tennessee Valley communities.

Instead, the Regional Planning Staff proposed that an entirely new approach be tried in Tennessee which had a strong state planning agency as well as a strong program of local planning assistance. It proposed that TVA and the Tennessee State Commission staffs join in a technical appraisal of the possible application of flood data to planning programs. The investigation was to include research into the types and form of flood information needed by state and local planning programs and how such data might be applied to community plans, zoning, subdivision regulations and capital improvement programs.[23]

These investigations resulted in TVA preparing reports designed to provide an understanding of the nature and magnitude of flood problems in specific communities or areas. The reports were to contain two sections: the first would provide an analysis of the history of floods in the community including information on flood heights, dates and frequency of occurrence, stream discharges, hydrographs of the rise and fall of flood waters, and maps showing the area covered by the greatest flood of record including stream cross-sections at important locations in the community. And finally it contained information on the size of flood that could reasonably be expected to occur in the future.

The second section of the report indicated the size of the floodway needed to pass floods downstream and the types of open space uses which might be permitted in such areas. It also delineated the flood fringe area on the edge of the floodway where most land uses could be permitted subject to control of floor elevations or flood proofing of structures.

Demonstrations in a few Tennessee communities were an immediate success and the program spread to communities throughout the Valley. By 1973 137 communities across the region had information on their flood situation and were incorporating this information into land-use plans. Thanks to the efforts of TVA many communities had included flood damage prevention measures into their zoning ordinances and subdivision regulations.

The demonstrations in the Valley had important implications nationally. In 1959 TVA submitted a report to Congress entitled 'A Program for Reducing the National Flood Damage Potential'. This report reviewed the experience of planning agencies in the seven valley states in using flood damage prevention measures to control development in areas subject to flood. The TVA report provided the basis for establishing the Task Force on Federal Flood Control Policy which in 1966 released a report entitled *A Unified National Program for Managing Flood Losses*.[24] This report, in turn, became the basis for the National Flood Insurance Program.

Looking back from the vantage point of the serious floods of 1993 in parts of the US it seems unfortunate that the national program that was adopted emphasized flood insurance and 20-year flood criteria rather than control of land use in flood plain areas where data for past floods indicated much greater magnitudes could be expected. Such controls could have been applied nation-wide since the Corps of Engineers had already started a program to provide communities throughout the nation with information on stream flow and flood data for use in their planning and development programs.

## Reservoir Shoreland Planning

In 1958 TVA decided to ask for appropriations for additional dams on the main river and its principal tributaries. The first structure, Melton Hill, was to be built on the Clinch River just downstream from Oak Ridge, Tennessee. It had long been the policy that once TVA submitted a recommendation to the Bureau of the Budget (Office of Management and Budget) and the recommendation was included in the President's budget, the recommendation could not be discussed publicly even with affected states and localities. As a result, TVA staff could not review the project with those state and local agencies affected by the project; neither could it enter into joint planning studies to develop a plan for the use of reservoir shoreland.

The Melton Hill Dam Project was first included in the 1958 budget. Congress, however, decided to postpone funding the project until the following year. This decision made it possible to try new approaches to reservoir shoreland planning. TVA started discussions of the proposed project with the state of Tennessee and affected local counties and communities. A technical planning group was set up with staff from the Tennessee State Planning Commission, the Tennessee Department of Conservation, the local planning agencies from Oak Ridge, Knoxville and Knox County, and Clinton, and TVA. Prior to the final authorization of the Melton Hill Project by Congress in the 1959 budget this group had developed a general plan for the reservoir shoreland including recommendations for highway relocation and potential sites for river terminals and industry, as well as areas suitable for commercial boat docks, public parks, public access areas, and residential areas. Such preimpoundment studies made it possible for both Oak Ridge and Clinton to establish port authorities and to construct marinas in the dry before the reservoir was filled. The plan had also been used to rezone shoreland areas.[25] It was the first time that it was possible to demonstrate the advantages to TVA of trying to merge project and shoreland planning. A similar approach was followed for Nickajack Reservoir before construction started on that project in 1964.

## The Tellico Project

Tellico Dam was the next of the main river projects for which TVA requested appropriations. This project, located at the confluence of the Little Tennessee and Tennessee rivers was first proposed in the 1936 plan for improvement of the Tennessee River system. It was conceived as a part of the Fort Loudoun and was originally called the Fort Loudoun Extension. The planning for this dam was a logical sequence in the development of TVA policy for merging reservoir project and shoreland planning.

About 1965 TVA started looking at the Fort Loudoun Extension, now with its new name Tellico, but with a different concept. The area along the Little Tennessee was largely open country with a few struggling small towns and very little economic activity. Located on the fringe of the Knoxville Metropolitan area there were already a few scattered and substandard subdivisions coming to the area.

An idea which attracted the attention and support of the TVA Board was to plan for a new town on the shores the new project using the Melton Hill and Nickajack experience to bring together project and shoreland planning. This varies from the view taken by Bruce Wheeler and Michael McDonald in their essay 'The 'New Mission' and the Tellico Project, 1945–70'.[26] They asserted that:

The building of the Tellico Dam on the Little Tennessee forced TVA to publicly articulate its new mission Although technically not a tributary but main-river project and hence not within the purview of OTAD, Tellico was the place where TVA believed it had to take a stand in defense of its new mission as the agency that changed nearly every aspect of life in the tributary areas.

This is a questionable point of view since the OTAD staff was never involved in the project. Moreover, Wheeler and McDonald claim that the land purchase policy used on Tellico was 'a reversal of a TVA policy since the building of Fort Loudoun Dam'. For each project the TVA Board established a land purchase policy and in the case of Fort Loudoun the restricted land purchases were primarily a function of the location of the project in a heavily populated urban area. As noted in the preceding section the land purchase policies for both Melton Hill and Nickajack permitted sizable purchases of land above the reservoir levels. So did the policies for Fort Patrick Henry (1953), Boone (1952) and South Holston Dams (1950). The land purchase policy for Fontana Reservoir (1944) permitted the acquisition of thousands of acres some of which were subsequently transferred to the National Park Service as an addition to the Great Smoky Mountains National Park.[27]

The Tellico project had long been a priority project of the Board and the Engineering Staff. It is probably true that then Chairman Wagner was a 'water man' who always supported dam building as an important part of the TVA mission. It is also true that the then standard cost-benefit analysis could not justify the Tellico Project. The idea of a new town when merged with the water project could strengthen project justification. Moreover, it was the Regional Studies Staff which had long supported reservoir shoreland planning that proposed the new town project and helped develop the new town plan.

To understand how this came about one need only look at some of the ideas that dominated professional planning thinking of that time. In the mid-60s there was much discussion on how to save the nation's cities and one of the alternatives was to promote the building of new towns. A new towns act had been passed by Congress but there was little support for the program. The British experience with the New Town Act of 1946 had been a success. Under both Labor and Conservative governments seventeen new towns had been built. In the United States the new town of Reston, Virginia was considered a success. The new town of Columbia, Maryland located midway between Washington and Baltimore was attracting attention and support from private developers and town planners.

Wolf Von Eckhart, writing in *Harpers Magazine* in 1965, expressed the then current views on new towns:

New towns would conserve precious open country side. Not every one would want to live in them, of course. But if we built just 350 new towns of 100,000 inhabitants each, they would house 35,000,000 people, or about half of the estimated twenty-year population increase.[28]

TVA's Regional Planning Staff must have shared these views. Beginning in the mid-1960s the staff started to build a concept for the emerging urban patterns in the Tennessee Valley. In part, this was a response to the 'growth center' concept used by the Appalachian Regional Commission and the Economic Development agency in allocating funds to development districts.

**Figure 3.5   Aubrey J. Wagner, TVA General Manager, 1954–1961, TVA Board 1961, Chairman, 1962–1978, a strong supporter of the regional development program but also a strong believer in water resource development. G. O. Wessenhauer, head of the TVA Power Program. He viewed the regional development program as complementary to the power program**

*Source*:   Patricia Bernard Ezzell, TVA Historian.

What the staff proposed was a concept of a system of economically and socially integrated cities centered on metropolitan centers and tied together by improved transportation and communication networks. This concept was viewed as a more dynamic framework for programs affecting population distribution and economic development. The concept was never fully accepted by the TVA Board even though the concept was described in the lead section in the 1972 TVA Annual Report as follows:

> We must help create a system of cities, towns, and villages with open space within and in between. Such a system is already emerging across the Valley region. The Valley's still evolving urbanization pattern has produced distinct groups of various-sized, independent but socially and economically interrelated communities. Like planets in mini-solar systems each group is centered on one of the Valley's larger metropolitan areas. Linked by improved transportation and communications networks, they can provide a framework for planned, orderly decentralization.

> The small and medium -sized cities and towns are the key elements in the concept. These are the areas of both the region's greatest population growth and its highest rate of growth, as people follow job trends into the countryside. Four decades ago, only a dozen communities in the Valley region fell within the 10,000 to 50,000 population range. Today, 42 are within this range. Only half of the region's counties contained any urban population as recently as 1950. Today that figure approaches 70 percent. . . .

> In the eastern part of the region, the Asheville, N. C. metropolitan area, and the Tri-Cities, Knoxville, and Chattanooga in east Tennessee each serve as the nucleus for a host of communities such as Morristown, Greeneville, Oak Ridge, Cleveland, and Athens. Huntsville's sphere of influence extends across north Alabama and northward into the lower Elk River valley. A similar system encompasses the central basin of middle Tennessee, reaching Clarksville, Murfreesboro, Columbia, and surrounding points. Paducah's force is felt across a section of western Kentucky and Tennessee. Memphis, the region's largest city, is the hub of a wide swath of geography sweeping from Jackson in west Tennessee through Tupelo and across north Mississippi.

> The areas thus defined are not static. There is flexible overlap and interconnection between them and, happily, plenty of open space within and wild land in between. But the pattern does provide a framework within which to plan and create desirable alternatives for living, alternatives which can touch the lives of virtually every citizen in the vast Tennessee Valley region.[29]

Although the concept was not fully accepted by the TVA Board it was used as a framework for several TVA urban projects, The first of these was the plan for the development of the Tellico reservoir shoreland. TVA proposed that a new town be built on the shores of Tellico Reservoir. It was to be named Timberlake after one of the first explorers to visit the Little Tennessee River area. In making the proposal, staff noted the sprawl extending west out of

Knoxville and the growing problems of providing urban services as well as the loss of prime agricultural land. It was proposed that the new community have its own economic base and a highway system to provide easy access to the employment opportunities in Knoxville.

TVA staff had visited new towns throughout the country. It discovered that a major problem for new town building was the assembly of land needed for a new town project. In its proposal for Timberlake TVA staff recommended a national demonstration of new town development that would pool public and private resources for quality urban growth. TVA's major contribution would be to assemble land for the new town. The proposal required entirely new procedures for both dam and reservoir planning.

The project investigations started with TVA and the Tennessee State Planning staffs discussing the proposal with Blount, Loudoun, and Monroe counties and the cities located in the area. These discussion lead to the organization of the Tellico Regional Planning Council. Membership in the Council included mayors and county executives, representatives of the city and county legislative bodies and local planning commissions within the three county area. The Tennessee State Planning Commission agreed to provide staff for the Council. TVA and its consultants would be responsible for the necessary technical studies including design but would consult with the Council on all findings and at all stages of plan formulation.

This view of the organization and purposes of the Tellico Regional Planning Council also varies from that expressed by Wheeler and McDonald who claim that the Council was organized because TVA ' . . . needed such a group to carry the Tellico banner'. In our view it was the kind of organization made up of planning commission members and representatives of local government that the Regional Studies Staff had traditionally worked with on reservoir shoreland planning studies throughout the Tennessee Valley.

As part of the initial proposal for the new town it was recommended that the traditional way of handling road and highway relocations be modified. Instead of merely reconnecting roads and highways flooded by the reservoir it was proposed that the allocated highway moneys be used to build, in cooperation with the state highway department, a regional highway system that would give easy access to the interstate system for both the new town and the communities in the Tellico area. The moneys would also be used to provide a major highway through the proposed new community to which residential, industrial, and recreation development could be connected. It also proposed a regional water supply system to serve shoreland developments and the entire three county Tellico planning area. Finally, a waste disposal system was to be designed that would protect water quality and assure that the overall environment of the area was protected.

The plan that finally emerged was for a new town to be located on the left bank of Tellico Reservoir. The right bank opposite the town would be kept as open space to protect the town's view of the lake. Approximately 16,000 acres of Tellico shoreland was to be made available for the town. The town was to be developed over a twenty year period and ultimately have a population of about 30,000. The industrial area was located upstream from the town near U.S. Highway 411 and the L. & N. Railroad. Once the reservoir was complete the industrial area would also have a nine-foot channel with connections to the inland waterway system. The upper end of the reservoir located between the Great Smoky Mountains National Park and the Cherokee National Forest would be used for a wide range of recreation services (Figure 3.6).[30]

When time came for TVA to implement the town and reservoir plan the TVA Board hesitated and repeatedly postponed action. In the meantime Congress repealed the new town legislation and funding from that source was no longer available for the project. The Boeing Company which had joined TVA in planning the new town and which had indicated an interest in building the community decided it could no longer continue as a part of the project.

Finally with the appointment of S. David Freeman as Chairman of the TVA Board there was no longer Board support for the Tellico Project. Congress, however, approved and provided funding for the project. Litigation over the environmental effects of the project produced many construction delays and the dam was not closed until November 29, 1979. The pros and cons of the environmental impact of the project have been reported and discussed in detail by the press and in professional publications and will not be reviewed here. Instead what happened in the area after the project was completed will be described.

Although land had been acquired and roads built as recommended in the reservoir shoreland plan there remained questions concerning the implementation of the plan. By 1980 development standards based on the plan concept were recommended to and accepted by the TVA Board. TVA then tried to find private developers to carry out the plan. A large share of the land in the aborted new town area was sold to a private firm, Cooper Communities, to build a retirement community. Although reasonably successful as a retirement area Cooper Communities has also focused its sales strategy on local residents working in the Tellico and the Knoxville Metropolitan areas, suggesting a shift to the kind of community envisioned in the original new town plan.

The Tellico Regional Planning Council was replaced by the Tellico Regional Development Agency (TRDA) which had powers to help implement the area development plan. TVA contracted with TRDA to manage land sales to manufacturing and service companies and sold the agency 1,100 acres of Tellico shoreland. In addition TRDA took responsibility for the development and management of public utilities and public access areas and for the enforcement of water quality and other environmental standards established for the reservoir and the reservoir shoreland.

By 1992 the reservoir plan was well on its way toward full implementation. Certainly area investments exceeded earlier projections. Over $161 million in new homes had been constructed on the reservoir shorelands. TRDA recruited

**Figure 3.6    Proposed Timberlake New Town, Model of Town Center,
TVA, 1976**

*Source*: TVA files.

industries which had constructed over 1.5 million square feet of manufacturing
and commercial space representing an investment of about $314 million and
providing employment for over 2,000 people. Unemployment in the three
county Tellico area fell from 12.6 percent in 1983 to 6.7 percent in 1992.
These developments also resulted in additional tax receipts of $4.2 million
for Loudoun County and $2.8 million for Monroe County.

To support these and other future developments TRDA built the Port of
Tellico, a waste water treatment plant, 2.3 miles of railroad, 2.8 miles of

industrial roads, 1,250 feet of gas lines, 11,120 feet of sewer lines, and 28,970 feet of water lines. These improvements provided important infrastructure for area growth. TRDA also has developed an innovative skills development program for the Tellico reservoir area. It built an 18,000 square foot training center including a fully equipped computer laboratory. The center is managed by the Tellico Educational Consortium composed of representatives of many community and technical colleges in the east Tennessee area.[31]

In 1993 TRDA sold 850 acres of land on the Loudoun-Monroe county line to Tellico Lake Properties, Inc. The plans for this property include 2,000 home sites around a 27 acre golf course. In addition Cooper Communities opened a new neighborhood containing 115 lakeside and lakeview homesites and announced the construction of a second golf course on its property. Such high quality development with built-in environmental safeguards is only possible when a group such as TRDA is dedicated to carrying out the plan for the new town on Tellico reservoir. This kind of dedication by the developer has been one of the most important ingredients for other successful new towns in the United States.

Looking back on the Timberlake/Tellico experience our conclusion is that it was more successful than could have been anticipated given changing TVA administrations and changing national political support for new towns. The new town that is emerging on the shores of Tellico Lake is very much like the town as originally planned by the TVA staff.

## North Alabama

On a somewhat different scale TVA tried another type of experiment in north Alabama to demonstrate what might be done to give some structure to urban development in a growing urban region. The population of this eleven county area grew from 460,000 in 1940 to 541,000 in 1960; an increase of 81,000 people most of which was added to the cities and suburbs of the area.

In north Alabama the growth of industry in the major urban centers of the 'Quad Cities' (Florence, Sheffield, Tuscumbia, and Muscle Shoals City), Decatur, and Huntsville and in such small centers as Athens, Scottsboro, and Guntersville, Alabama and in Pulaski and Fayetteville in adjoining portions of Tennessee had resulted in an extremely low density urban growth throughout the region but outside the cities. This population is concentrated along the major highways and the country roads and is possible because good transportation has permitted these extra-urban residents to commute to full time jobs in or near the urban centers. A second factor permitting a scattering of population was the policy of the Department of Agriculture to encourage and help finance water systems in places distant from the urban centers. This questionable policy is not related to any concept for encouraging sound and stable urban growth.

North Alabama, particularly in the Highland Rim section extending along both sides of the Tennessee River contains some of the best agricultural land in the entire Tennessee Valley. The low density extra-urban population has tended to break up farms and decrease the agricultural productivity of the region. The low-density urban sprawl represented by this population distribution pattern has resulted in higher public costs for such essential services as education, water, waste disposal, health, and police protection. Population projections indicate that this segment of the population will continue to grow rapidly thus adding to these problems.

## Institutional Resources

The expanded institutional base which TVA had encouraged during the first twenty years and the continued growth of these resources during the 1950s and 1960s provided a starting point for TVA activity in the area. Practically all of the cities and even some of the counties had active planning agencies with technical assistance provided by the Alabama State Planning Board. The state planning board also joined with localities in an industrial development program. Many cities and counties also had recreation commissions.

The entire North Alabama Region was included within three development districts: the Muscle Shoals Council of Governments (MSCOG) in the west; the North-central Regional Council of Governments (NARCOG) in central North Alabama; and the Top of Alabama Regional Council of Governments (TARCOG) in the east. In addition TAD worked with tributary development associations in the Bear Creek and Elk River watersheds, both on the fringes of the urbanizing portion of North Alabama. These institutions which continue to exist have provided the base for a broad urban development program in the area.

## The Elkmont Experiment

In 1969 TVA and state, regional, and local planning agencies started to study the problem of strip development along major highways in the area. One idea that emerged from these discussions was a proposal to build free standing villages in the open countryside. These villages would offer a full range of public services and would be an alternative unplanned strip development. The Elk River Development Association became interested in the village concept and in 1969 asked TVA to assist in preparing a proposal that would help preserve the open country atmosphere of the lower Elk River while providing a high level of service for new housing needed in the area. Three counties – Lincoln and Giles in Tennessee and Limestone in Alabama were within the lower Elk river watershed. Limestone County was also within the jurisdiction of TARCOG. After two years of study and meetings with local agencies and

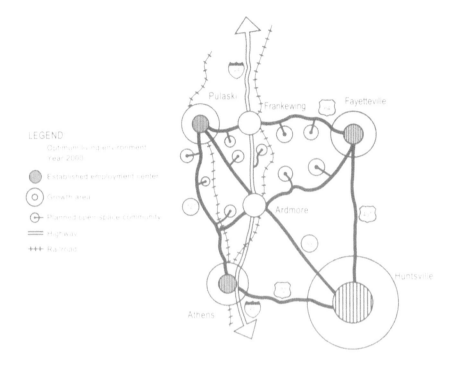

**Figure 3.7 Planning for the Elk River Area, Middle Tennessee and North Alabama, based on the concept of planned open-space communities**

*Source:* TVA Annual Report, 1972.

officials and the general public a plan began to emerge which called for nine villages – three in each county (Figure 3.7). Each village would contain about 1,500 acres and would accommodate 3,000 to 4,000 people.

A key to the concept was to establish a revolving fund for construction of the villages with funding of $4.7 million for the initial village. Revenue from the sale of land and excess income from utilities were to be paid back into a fund which would be used for a second village.[32]

The site selected for the first village was immediately west of the Town of Elkmont, Limestone County, Alabama. The general area is typical of much of North Alabama. Although the dominant land use is agriculture there is a large and growing number of non-farm people commuting to employment centers in and near the region's cities. Elkmont and the site of the proposed village is only about 10 miles south of Athens, Alabama and only 3 miles west of I-65, which gives good access to the larger employment centers to the south and east.

The first village was planned for a 1,450 acre site to be divided into 900 acres of land for development and 500 acres of open space mostly on the periphery of the village. The developed area would have 1,100 single family home sites most of which were located on lots less than one acre in size. The remainder of the developed area was to be in mini-farms of about four acres. The plan also provided for a small industrial park south of Elkmont and adjacent to the railroad. There was also to be an historical park and recreation area at the site of the Civil War Battle of Sulphur Trestle.

The first village was built by a development corporation sponsored by the Elk River Development Association and funded by a congressional appropriation. The Alabama State Highway Department improved the three-mile access to I-65 and the Bureau of Recreation provided funds to improve the recreation support facilities of the Town of Elkmont. The project was modestly successful but changing TVA interests and management did not produce the kind of continuing commitment and follow-up which was required for the full success of a project and concept of this type. The Elkmont Village was the only one of the proposed nine villages built. Without support from the TVA Board and the new top management the concept was lost as TVA searched for new program directions. It is interesting, however that the revolving fund remained available until the early 1990s when Congress shifted the moneys to other uses.

## Townlift Program

An important adjunct to the planned village proposal was an expanded program of urban and industrial planning assistance to cities and towns in the region. By the mid-1960s all of the Valley states had local planning assistance programs staffed with trained city and regional planners. TVA help was no longer needed for general planning assistance to cities and counties. TVA did start a 'Townlift' program which was aimed at the revitalization of small- and medium-sized communities so that they could compete with the larger metropolitan cities and in the process build a system of cities as an alternative to the ever expanding metropolitan centers.

The program used the broad range of technical skills available within TVA to help cities and counties find ways to solve specific problems. Multidisciplinary teams were formed composed of the mix of professionals needed for a particular problem. As needed, city and regional planners, economists, public finance experts, architects, engineers, and recreation planners served on these teams. The teams helped on specific community development problems such as downtown revitalization, industrial site analysis and site planning, financial planning, and recreation site and program planning. Townlift was channeled through state planning agencies to local planning agencies.

In North Alabama townlift assistance was made available to almost every city including Florence, Sheffield, Tuscumbia, Decatur, Huntsville, Athens, and Guntersville. In Decatur, for example, TVA worked with the state and local planning agencies to assure that land designated for industrial use did not interfere with the operation of the adjoining Wheeler National Wildlife Refuge. It also helped to relate a proposed new state dock to the waterfront industrial areas in the Decatur area.

On a somewhat different scale TVA provided planning and engineering assistance to the Muscle Shoals and Top of Alabama regional development agencies (MSCOG and TARCOG) in studies of possible regional (multi-county) water and sewer systems. Such studies looked at present and future residential and industrial areas and projected the water and waste water demands. The end product was a regional framework for future expansion of such facilities by counties and municipalities in the region; a prerequisite for water and sewer grants from the Department of Housing and Urban Development and the Environmental Protection Agency.

Townlift assistance was not confined to north Alabama. In 1971, 35 communities in the Tennessee Valley Region were included in the program. Some examples are the downtown improvement and beautification program for Waynesville, North Carolina. Here the Board of Aldermen for the city pledged $40,000 annually for as long as needed to complete implementation of the project. In Cumberland City and Erin, Tennessee near the TVA Cumberland Steam Plant the program helped with housing and community service problems created by the construction of the plant. In Big Stone Gap, Virginia studies were made of ways to alleviate flood problems in the context of overall redevelopment needs. In Fort Oglethorpe, Georgia a townlift team helped the city with studies of new interstate access and highway by-pass and a process for dealing with problems of strip highway development.

## Regional Highway Connections

The growing cities of north Alabama had good access to the metropolitan areas to the north (Nashville and Louisville) and to the south (Birmingham) but lacked good and easy access to major centers to the west (Memphis) and to the east (Atlanta) To look at this problem the regional planning agencies in north Alabama, North Mississippi, and northwest Georgia with help from TVA cooperated in a study to determine the need for and the feasibility of constructing a high speed, limited access highway through the region to connect with Memphis and Atlanta.

Although the study established the feasibility of the route, funding could not be found and the project was not built. However, some 20 years later in 1993 the project was again actively proposed by the regional planning agencies, the local governments, and the chambers of commerce of north Alabama.

## Summary: An Overview of TVA's Regional Development Activities in its Second Twenty Years – 1953–1973

Although there was an active regional development program during TVA's second twenty years the program did not have the focus of the preceding twenty years. In general the activities associated with the earlier period continued. Several significant new activities were initiated such as the flood damage prevention program which provided a demonstration of possible solutions to a growing national problem. Also new was the tributary development program which attempted to provide a means for pulling the TVA development activities together. Then there were the urban programs. But all of these did not fit into the kind of coherent and identifiable overall program concept which institutional building had provided during the first twenty years. Certainly, the TAD program which was to have been the centerpiece of TVA's regional development activities in the second twenty years did not give a clear cut direction to staff activities.

This lack of direction was due in part to TVA's failure to recognize the economic and demographic changes taking place in the Valley and in the South. This is evident from the choice of tributary areas as the focus for TVA regional development activities during this period. From the experience of the first twenty years TVA should have understood that the watershed area definitions used by TAD, while suitable for water resource planning, were not appropriate planning regions for dealing with the problems growing out of the new conditions in the larger region.

This kind of problem is one which regional agencies in general have faced and few have been able to solve. The fact that TVA was a regional agency, as Hargrove noted, 'suspended at mid-point between the national government in Washington and State and local government'[33] was also a part of the problem. Having completed the specific regional tasks outlined in the TVA Act, there remained the problem of identifying new and specific tasks that were regional in scope. Although Hargrove raised this issue, few of TVA's friends or critics or even top TVA management recognized its significance to planning the long term regional program.

The continued growth of the power program and its dominance of regional development planning certainly limited the range of options open for consideration of regional needs. Toward the end of the second twenty years the problems within the power program were so large and complex that they were to dominate TVA Board attention, thus limiting the kind of attention the overall regional program needed at this crucial time. Too, the managers of the power program and the TVA Board had always opposed opening the Act for changes, thus making it impossible for TVA to consider new and needed regional programs.

Another staff change could have had a significant effect on the overall TVA regional program. From the beginning TVA had an independent staff concerned with regional and urban affairs: first, the Division of Land

Planning and Housing, which later became the Division of Regional Studies, and then the Government Relations and Economics Staff within the Office of the General Manager. This staff had always been an independent voice in TVA representing urban interests and intergovernmental cooperation as a basis for the regional program. In 1968 this staff was joined with the navigation division to form the new Division of Navigation and Regional Studies. In the process the regional studies group was divided into separate staffs; regional and city planning, economic research, and governmental relations. The result was that there was no longer an independent voice to represent before the Board a totality of areal planning interests. The result was a further weakening of the already small influence this staff was able to exert on TVA policy options.

This change was symptomatic of a problem which had long plagued TVA management: the almost paranoid belief that any form of central planning would be harmful to both the operation and image of the agency. Lilienthal rejected and seemed to dislike the word 'planning'. He tolerated but did not actively support Earl Draper's efforts to organize the Regional Planning Council as a way of pulling together the TVA divisions concerned with resource development. Yet the original board, in spite of the tensions created by internal conflicts and external challenges, did establish policies which were to guide effectively TVA regional development operations for two decades. These policies constituted a regional plan even though the Board and major staff would not recognize it as such. Moreover it was an achievement which future boards were not able to accomplish.

Hargrove notes that in the early 1970s the then General Manager Lynn Seeber became concerned that TVA Chairman Wagner depended too much on his own knowledge and experience and his personal relationships with division directors to reach major policy decisions. Seeber believed there was to little systematic analysis of possible policy options. To meet this problem Seeber set up a small central planning staff within the general managers office. The economist who directed the staff recalls that Wagner had little interest in the work of the planning staff. Instead, Wagner had confidence in his division directors and believed they were already making the best recommendations possible. Hargrove noted that Wagner confirmed this view by stating in an interview 'that it is a mistake to have central planners tell people what to do because they are not responsible for results'.[34] This un-robust, fearful view of planning unfortunately dominated Board thinking at this time.

This was not a new experience in TVA. As early as 1953 the Government Relations and Economics Staff in the Office of the General Manager began quarterly reviews with the Board of the social and economic changes and conditions in the Valley including income, changes in population distribution, and major economic trends particularly those relating to employment patterns. In 1966 Peter Stern was brought in to head this staff in the hope that he could find a way to use such staff skills to help the Board in considering options in policy decisions. All of these efforts failed because of the attitude expressed above by Wagner. The Board did not feel comfortable requiring the divisions responsible for planning and operating the individual TVA programs to defend those programs on the basis of data on economic and demographic conditions in the Valley which had been developed by an central TVA planning staff. If the Board had looked at these independent data projections it might have been able to avert the severe problems that resulted from acceptance of the high-side projections of the Power Staff for continuing increases in power demand.

Finally in the mid-1970s there was a complete change in TVA management. In this process TVA lost sight of its administrative and policy history. By then it was clear that the Aubrey Wagner view of regional development was no longer adequate and changes were needed. Under the new Chairman, S. David Freeman, who succeeded Wagner, TVA was never to be the same.

Avery Leiserson in his excellent essay 'Administrative Management and Political Accountability' has made an observation that was to be particularly germane to TVA's experience in its third twenty years and almost a prophecy of what was to happen in the twenty years from 1973 to 1994:

> The Authority will be fortunate if it does not . . .[suffer] the fate of many independent corporations and government agencies whose statutory authority has been so legislatively diluted and their policy-making structure so infiltrated by persons representing partial group viewpoints, that they have lost most of their effective powers of initiative and discretion to take a comprehensive view of the public interest'.[35]

## Notes

1. Tennessee Valley Authority, *Watts Bar Steam Plant Studies*, Knoxville Tennessee, Sept. 1949, p.1
2. TVA *Annual Report* for fiscal year ending June 30, 1953
3. TVA *Annual Report* for fiscal year ending June 30, 1954
4. *Knoxville News-Sentinel*, August 19, 1953, p.1–2
5. Aaron Wildavsky, *Dixon-Yates: A Study in Power Politics*, Greenwood Press, Westport, CT, 1976, p.10
6. Roscoe Martin, *TVA, The First 20 Years, A Staff Report* University of Alabama Press and the University of Tennessee Press. 1956, p. 266–267
7. Wildavsky 1976, p.313
8. TVA *Annual Report* for the fiscal year ending June 30, 1951, p. 50–51
9. TVA *Annual Report* for the fiscal year ending June 30, 1952, p.70
10. Tennessee Valley Authority, *Working with Areas of Special Need with Examples from the Beech River Watershed*, June 1953 p. 3
11. Melvin R. Levin, 'The Big Regions', *Journal of the American Institute of Planners* March 1968, p. 69
12. TVA *Annual Report* for fiscal year ending June 30, 1961, p.1
13. See in particular the *Louisville Courier-Journal* for Sept. 4 through 9, 1961 and the *Nashville Tennessean* for July 16, 1961
14. Memorandum A. J. Wagner, TVA General Manager to TVA Board of Directors, dated April 20, 1958 entitled *Upper French Broad River Flood Problem*

15  Memorandum A. J. Wagner to Geo. K. Leonard, TVA Chief Engineer, dated May 16, 1958, entitled *Upper French Broad Flood Problem*

16  Memorandum A. J. Gray to L. L. Durisch, dated Nov. 5, 1958, entitled *Upper French Broad Flood Control Project*

17  Informal memorandum A. J. Wagner to L. L. Durisch, dated Nov. 13, 1958

18  Memorandum A. J. Gray to Peter M. Stern, Director of Regional Studies, dated April 8, 1966, entitled *North Carolina – Development of the Upper French Broad*

19  Memorandum A. J. Gray to L. L. Durisch, dated Nov. 19, 1962 entitled *An Approach to Urban-Industrial Problems in the Tennessee Valley* with cover transmittal from L. L. Durisch dated Nov. 20, 1962 to Louis Van Mol, TVA General Manager

20  Aelred J. Gray, 'Urbanization: A Fact – A Challenge' *The Tennessee Planner*, April 1957, pp. 145–155

21  Memorandum C. E. Blee, Chief Engineer to George F. Gant, TVA General Manager, dated Mar. 16, 1951, entitled *Major Flood Problems in the Tennessee River Basin – Report of September 8, 1950*

22  Memorandum L. L. Durisch, Chief of Government Research Staff to William J. Hayes, Acting Director of Regional Planning Studies, dated Nov. 28, 1951, entitled *Local Flood Damage Control Problems and the Work of State and Local Planning Agencies*

23  Aelred J. Gray, 'Planning for Local Flood Damage Prevention' *Journal of the American Institute of Planners*, Winter 1956, pp. 11–16; Aelred J. Gray , 'Communities and Floods' National Civic Review March 1961, pp. 134–138

24  House Doc. No. 464, 89th Congress, 2nd Session, U.S. Printing Office, Washington, 1966

25  Aubrey J. Wagner, 'Natural Resources – A Challenge for Planning' *The Tennessee Planner*, March 1965, pp. 71–72; 'Melton Hill Reservoir: Comprehensive Plan for Land Use Development' Tennessee State Planning Commission, Publication #310, Dec. 1960

26  Edwin C. Hargrove and Paul K. Conkin, (eds.), *TVA Fifty Years of Grass-Roots Bureaucracy*, Univ. of Illinois Press, 1983; pp. 168

27  Howard K. Menhinick, 'Supreme Court Decision Assures Addition to Great Smoky Mountains National Park', *Planning and Civic Comment* April 1946, pp. 46–48

28  Wolf Von Eckardt, 'The Case for Building 350 New Towns' *Harper's Magazine*, Dec. 1965, pp. 85–96

29  TVA 1972 *Annual Report*, pp. 9 & 10

30  Timberlake New Community General Land Use Plan, Tennessee Valley Authority, April 1976, 33 pp.

31  Tellico Reservoir Development Agency, *Decade of Progress* Sept. 1992, 6 pp.

32  Environmental Statement, Elkmont Rural Village, Tennessee Valley Authority, Dec. 16, 1974

33  Edwin C. Hargrove and Paul K. Conkin, (eds.), *TVA Fifty Years of Grass-Roots Bureaucracy*, Univ. of Illinois Press, 1983; pp. 168 and 179

34  Edwin C. Hargrove, 'The Task of Leadership: The Board Chairmen' in *TVA: Fifty Years of Grass-Roots Bureaucracy* University of Illinois Press, Urbana and Chicago, 1983, p. 111

35  Avery Leiserson, 'Administrative Management and Political Accountability' in *TVA: Fifty Years of Grass-Roots Bureaucracy* University of Illinois Press, Urbana and Chicago, 1983 p. 143

# The Rise of the Corporate Image and the Managerial Elite: Effects on the Regional Development Program, 1973–1994

Regional agencies such as TVA which are outside the traditional governmental structure and which have specific regional responsibilities periodically face the challenge of change from both internal and external forces. While TVA tried to meet this challenge in its second twenty years it was modestly successful at best.

The challenges of the third twenty years were to be even more formidable. First, the TVA power program was confronted with problems which at times seemed almost unmanageable. These problems not only affected power operations but the financial stability of TVA.

Second, overall TVA operations, such as the location of power generation stations and transmission lines, air pollution, the problems of reservoir maintenance and management, were faced with changing public concerns over the effects of these operations on the environment. In this situation operations were becoming so important that the regional development program became a short term defense of operations rather than a long term response to regional needs. As many students of regional development programs have observed, a short term response to problems may run counter to long range regional or area needs.

Third, a new breed of managers was to replace the heads of technical staffs and dominate this period in TVA history. The regional development program was the major casualty of this change. The new managers failed to understand the lessons of the past because they had little or no training or experience in regional development. Neither did they understand the position of a regional agency relative to the traditional governmental units. Moreover most of these managers were recruited from private industry and viewed TVA as a utility corporation rather than a government agency with a public purpose. As a result the managers tended to internalize the regional development program rather than continue to develop cooperative regional programs with state and local governments; a practice which had served TVA so well in the post. As a result the regional development program tended to be perceived by the new managers as a problem in managing TVA operations such as the TVA reservoir shoreland properties or as an adjunct and support for the power program.

## Nuclear Power and TVA Finances

The environmental problems which TVA was having at its coal-fired plants caused it to look for alternative sources of power. Aubrey Wagner as Chairman of the TVA Board seemed to believe that most of the TVA problems at these steam plants grew out of the unreasonable demands of the environmentalists. As an engineer the nuclear plants seemed to Wagner to provide a good solution. At the time such plants were considered safe and pollution free. From this point of view the decision to start a nuclear plant program seemed rational.

TVA nuclear power program started slowly with the 1966 authorization of three units, each with a capacity of 1,152,000 kilowatts, at Browns Ferry near Decatur, Alabama. This was followed by the 1969 authorization of two units, each 1,221,000 kilowatts, at the Sequoyah site near Soddy-Daisy, Tennessee and the 1971 authorization of 2 units, each 1,270,000 kilowatts, at Watts Bar. The first Browns Ferry unit became operational in 1974 and construction was proceeding on schedule. The nuclear program seemed to hold the promise for future power generation needs.

TVA projections of system power needs suggested a continuing and growing power demand. Suddenly four additional nuclear plants were approved by the TVA Board: Hartsville north of Nashville; Yellow Creek in northeast Mississippi; Bellefonte in northeast Alabama; and Phipps Bend in upper east Tennessee. These plants would provide all of the energy needed in the TVA service area well into the 21st century; energy that was cheap and environmentally clean. Given the poor track record for reliability of long term economic projections and the new and untested technology involved, the basis for the TVA Board decision to start such a large nuclear program is not clear from the record.

A series of events were soon to throw TVA into turmoil. The first was the 1973 oil embargo which forced the United States to look at its energy consumption and to start conservation measures as a defense against dependence on foreign oil. TVA itself supported the power conservation program encouraging energy efficiency in homes and businesses throughout the Tennessee Valley region. Within five years this program reduced power demand substantially with the result that earlier TVA power projections proved much too high.

Then on March 22, 1975 a fire at the Browns Ferry plant forced a shut down of one unit. Coming at a time when there was increasing concern over the safety of nuclear power generating stations this mishap was probably the starting point for the whole TVA nuclear program to come unglued. Certainly safety concerns started the escalation of construction costs which eventually forced increases in power rates and a rising TVA debt that was to require changes in the entire power program.

## Changing TVA Management

A. J. Wagner retired in 1977 and S. David Freeman was appointed by President Carter to serve as chairman of the TVA Board. Serving with him on the Board were Richard M. Freeman (appointed in 1978) and Robert N. Clement (appointed in 1979) This board took the position that it was necessary to have an adequate supply of energy with quality economic growth. To do this the board concluded it was necessary for TVA to set a standard for the efficient and safe operation of nuclear reactors; use the newest of technologies to assure TVA coal fired generating stations did not contribute to air pollution; and give greater attention to energy conservation and alternative renewable energy sources.[1]

Under the leadership of S. David Freeman TVA started to defer and eventually cancel the Hartsville, Yellow Creek, and Phipps Bend nuclear plants. Bellefonte was deferred but recently a decision was made to resume engineering work at the site in March 1993. By 1994 the fate of this plant was still being considered by the TVA Board. Finally in March 1985 TVA decided because of safety concerns to leave the Browns Ferry units out of service and in August 1985 the Sequoyah was shut down for similar reasons.

Freeman also was a leader in expanding TVA's energy conservation programs as a means of meeting future energy needs through demand reduction and the use of alternative energy sources. He did this by offering financial incentives to power users and by offering free residential and commercial energy audits, low cost loans for home insulation, and the use of alternative energy sources such as solar, wood, passive building construction, and co-generation.

While the conservation program was a success, in retrospect it is unfortunate that steps were not taken to demonstrate that smaller and uniformly designed nuclear plants could reduce costs and increase operational and maintenance efficiency and safety. Instead TVA decided to continue its series of one-of-a kind plants which may have been the only decision possible considering the commitments in money and time which already had been made on existing and proposed plants.

Freeman also must be credited with stopping TVA's resistance to environmental regulation and with investments to clean up stack emissions from its coal-fired steam plants During the 1960s TVA had not given adequate attention to pollution from its coal-fired steam plants and gained the reputation as a major polluter. TVA even joined with other privately owned utilities in opposing environmental and nuclear safety regulations. Freeman changed this. He negotiated and signed an agreement with EPA to clean up

TVA power plants – an agreement his predecessor, Aubrey Wagner refused to sign.

The effects of these events on overall TVA operations are well documented. The problems of the power program were to so dominate TVA actions that the overall regional development program could not compete for the kind of serious attention it required and it thus become an insignificant part of the TVA program. Also, monetarily it was becoming such a small part of the total agency budget which added to its insignificance. Equally significant for the future of the TVA regional development program was the 1977 retirement of Aubrey Wagner as Chairman of the TVA and the appointment of S. David Freeman to that position.

**Figure 4.1   S. David Freeman, a lawyer and engineer who was appointed by President Carter to the TVA Board in 1977. He was Chairman from 1978–1984. His major accomplishments were the cancellation of several nuclear plants and the development of a sound energy conservation program**

*Source*: Patricia Bernard Ezzell, TVA Historian.

Aubrey Wagner had come up through the ranks. He understood the significance of the regional development program and how it had been built upon a system of regional cooperation between states, localities, and TVA. If he failed to understand the economic and demographic changes taking place in the Tennessee Valley Region and the South he did recognize that as a regional agency TVA had to have the support of the Valley states and their leaders to achieve regional development plans. With the retirement of Aubrey Wagner TVA seemed to lose its historical memory and more importantly its sense of mission in the Tennessee Valley Region.

Some TVA employees claim that Freeman came to TVA with a 'Washington mind-set' rather than a sense of TVA as a regional agency. They also claim he did not understand that his constituency was the leaders and people of the Tennessee Valley rather than those whose views were currently in vogue in Washington. Many felt this led to dependence on political decisions rather than technical competence. The bureaucracy he built seemed to reflect these views and resulted in a reduced role for TVA's middle management which had been one of TVA's strengths over the years.

In his 1979 interview with *Newsweek* reporters Freeman stated '. . . he is trying to revitalize an agency that lost its sense of social mission over the last twenty years. TVA is not just another utility. Let's go back to our roots' The record suggests that Freeman wanted to put his own stamp on the agency and in doing this his association with environmental groups certainly colored his views. For example Freeman did not support the Tellico Project until Congress forced the issue by giving final approval and funds for the project.[2]

**The 1979 Reorganization**

To help him put his own stamp on the agency Freeman hired a consultant to find new personnel and prepare plans for staff reorganization. This job was eventually given to an existing staff member. The result was the first of many reorganizations, all without a clearly articulated mission or purpose, which was to create employee morale problems for TVA on into the 1990s.

The record is not clear as to who was considered for management jobs with TVA but the 'short list' did include Dr. Sharlene Hirsch. Dr. Hirsch had a Ph. D. degree in management from Harvard University. She had worked for a non-profit planning firm in San Francisco which had national and international projects in education, manpower planning, aging, and related fields. While on leave from this organization Hirsch had served as Coordinator for Human Resources Policy Planning for the Carter-Mondale transition team.[3] She had also worked for a Senate committee where she had met David Freeman.

In her interview Dr. Hirsch impressed the TVA Board and she was hired to work on community and economic development programs. Dr. Hirsch, while on the Carter-Mondale transition team, learned that a Department of Community Development was being discussed. She proposed that TVA establish an Office of Community Development as a demonstration of what might be tried at the federal level.[4]

At the same time there were major changes in TVA top management. Lynn Seeber resigned as General Manager in April 1978 and Dr. Leon Ring was appointed to that position in November 1978. John Stewart was brought in to head the planning and budget staff, a position which had been held by John S. Barron who had been appointed to that position by Seeber with the approval of Wagner. Stewart had a Ph.D. in Political Science from the University of Chicago and had worked as Administrative Assistant to Vice President Hubert Humphrey. He also had worked with Humphrey in the Senate where he served as Staff Director for the Senate Subcommittee on Science, Technology, and Space. It was during this period of service in Washington that Stewart first met Freeman.

In February 1979 a reorganization plan was announced. This plan grouped TVA staff into six main offices; three existing ones Agriculture and Chemical Development, Power, and Engineering Design and Construction. Three new offices were created: Natural Resources headed by Dr. Thomas H. Ripley; Community Development headed by Dr. Hirsch; and Management Services headed by William F. Willis.

Ripley had been with TVA for several years and in general continued the forestry and fish and wildlife aspects of the TVA regional development program all of which had been well established over the years. Some staff members believe he gave greater emphasis to the fish and wildlife aspects of the programs but these were slight and almost unnoticed changes. Ripley was also positioned to champion the emerging environmental movement within TVA which as is clear from the record was Freeman's real agenda. Dr. Hirsch however undertook a complete reorganization of TVA's community and economic development and TAD programs. After her first two weeks on the job Dr. Hirsch submitted a long memorandum to the then two Board Members, S. David Freeman and Richard M. Freeman, outlining her ideas for the new Office of Community Development.[5]

A review of this lengthy memorandum suggests that Dr. Hirsch did not understand the informal process of staff coordination nor the freedom given to the individual staffs to carry out their responsibilities within the framework of the overall TVA mission established by the TVA Board. Neither did she understand how TVA's regional resource development programs were interrelated and coordinated. Instead she saw them as a miscellaneous system of services; in her words 'a Red Cross truck service model'.

The alternative she proposed was a top heavy organization with an Executive Office of the Manager of Community Development that included an assistant manager, and offices of Intergovernmental Relations and Community Relations, Personnel and Staff Development, Community Program Information, and Program Evaluation plus an Assistant to the Manager for Program Affairs.[6] At the operating level there were to be three divisions: Regional Studies, Commerce, and Community Services. The

Figure 4.2    John Stewart, Ph.D. in Political Science, University
of Chicago; Administrative Assistant to Sen. Hubert
Humphrey. Appointed by Chairman S. David Freeman
to head the TVA Planning and Budget Staff, 1978–1988.
Resources Group, 1988–1992. Stewart's report, *Strategies
for the 1980s, A Statement of Corporate Purpose and
Direction,* was the first to emphasize TVA as a corporation
rather than a federal agency, a change that was to dominate
TVA thinking in the late 1980s and 1990s

*Source*: TVA Staff Photo.

Community Services Division had nine branches: Community Organizers, Floodplain Program, Public Technology, Community Planning, Health Programs, Local Economic Development, Manpower Programs, Human Services Program, and Mitigation Program. The work of all three operating divisions was to be integrated by four area offices each headed by a Regional Manager. This organizational structure took away the managerial authority of the operating staff chiefs and was to produce conflicts in program direction and purpose.

The memorandum did not include a statement of the community development mission, the place of the TVA community development activities in the region or how TVA activities in this field related to similar activities being carried on by states and localities. The goals of the Office of Community Development stated in the memorandum were to build public and political support for TVA in the Valley and in Congress and to help the poor. Considering TVA's weak legislative base in Sections 22 and 23 the reorganization did not address how TVA's Office of Community Development was to be a demonstration for a similar type of organization at the Federal level. TVA funding of community projects was proposed but the purpose or basis for such a proposal was not outlined.

This reorganization was the start for building a top-heavy bureaucracy within TVA. Moreover, since neither S. David Freeman or Dr. Hirsch were able to articulate either a resource or community development mission for TVA, staff became confused and demoralized. This was the beginning of a major problem that was to plague TVA in the years ahead.

Dr. Hirsch did not seem to understand her place in the TVA organization. She apparently took the position that since the Board appointed her she should report directly to that body. One immediate result was that General Manager Dr. Leon Ring resigned seven months after taking the job. He cited as the reason for his resignation that the then two TVA Board members, Chairman S. David Freeman and Richard M. Freeman would deal directly with office managers under him leaving him without a leadership role in the agency. William F. Willis was appointed General Manager. Willis had headed the Office of Management Services under the 1979 reorganization plan.

Dr. Hirsch sparked a problem for the Community Deveopment Division when, in a June 25, 1979 speech in Chattanooga, she stated her strong opposition to highway strip commercial development and indicated that TVA would purchase land along highways to prevent such development. Both Representative Marilyn Lloyd and Senator Jim Sasser wrote to the TVA Board Chairman Freeman protesting the speech. Senator Sasser in a letter dated July 16 asked for 'a detailed justification of the operation of Community Development, along with a fact sheet outlining the mix of finances for this office through appropriated funds and power and nonpower proceeds'.[7]

When it became time for making a budget request Dr. Hirsch made what appeared to be an inflated request. When the TVA Board did not honor her request she resigned effective October 31, 1979.[8] The damage to the TVA community and economic development program and to TVA employee morale had been done. Two different managers tried to make the Office of Community Development an effective operation but were not successful. By this time it was becoming increasingly clear that a redefinition of the TVA overall resource development mission was needed.

**Figure 4.3    William F. Willis, TVA Engineer, TVA General Manager, 1979–1991. President, Resources Group, 1991–1993. A strong supporter of the regional development program, he tried to articulate a new TVA mission but TVA board members did not move on his suggestions**

*Source*: TVA Staff Photo.

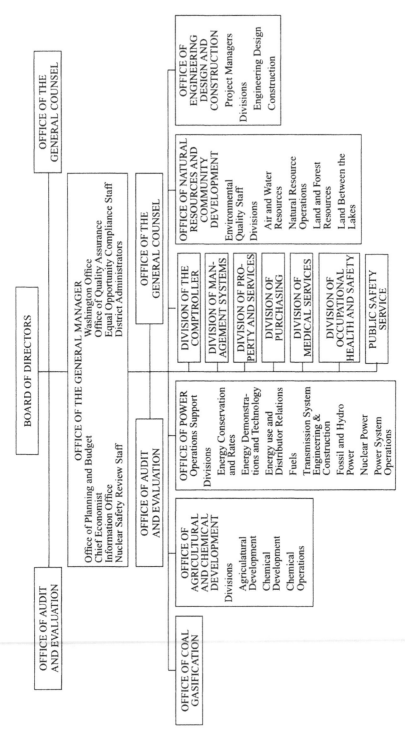

**Figure 4.4  Organization of the Tennessee Valley Authority, December 1983**

*Source:* Tennessee Valley Authority.

In 1979 TVA commissioned 11 university professors to help decide what TVA should be doing in the 1980s. The papers prepared by the university people between June and October 1979 were studies of: TVA's procurement policies; techniques TVA might use to develop closer working relationships with Tennessee Valley residents; TVA's work in urban development; rural development initiatives in the Tennessee Valley; how TVA can best deliver its programs to the public; TVA's relations with other government institutions; development of strategic planning alternatives for TVA; TVA and quality growth management; and the relative benefits and costs of alternative electric rate structures.[8]

This work was part of a larger program started by John Stewart, head of TVA's Office of Planning and Budget, to define the purpose and role of TVA over the next decade. The draft report of this office, *Strategies for the 1980s: A TVA Statement of Corporate Purpose and Direction*, was circulated to heads of major TVA divisions and to the TVA Board in 1980. While advocating change the report proposed goals similar to those of the past. For example, it proposed as an objective an adequate supply of power at the lowest possible cost. Another goal was that TVA should continue to support the efforts of state and local organizations to attract high-wage industry to the valley. There were similar general statements on development of valley resources and manpower.[9]

But the report did not deal specifically with the place of TVA as a regional agency in the Valley; how it would relate to states and localities as well as to other federal programs. It certainly did not recognize that the earlier TVA regional development program had been tied to TVA's specific responsibility as a regional agency to improve the river for flood control, navigation, and power generation. What were to be TVA's new specific regional responsibilities? How would TVA relate its powers to those it shared with states and localities? Equally important it did not seem to recognize that the Tennessee Valley and the South were urban regions and required new programs and approaches.

The failure of the TVA Board to articulate the new TVA mission was at the core of what was to become a major TVA problem. The publication *Inside TVA* described the problem in its December 15, 1981 issue:

> For most of the agency's life, TVA employees have known where the organization was headed. Then when TVA entered the nuclear power age in the 1970s and the size of the agency increased, veterans had to make room for newcomers who weren't as familiar with the history of TVA, and new programs were added to respond to the needs of the residents of the Valley. As the number of employees grew, so did the problems of communication. The rapid changes brought with them an increase in the complexity of communications between offices and divisions and at times left employees out in the cold about overall corporate direction.[10]

What this statement did not say was that during the Freeman years a huge bureaucracy had evolved with each group protecting its turf and the Board unwilling or unable to spell out an effective new direction for TVA that could be understood by all. Discussions with TVA employees during that time indicates that many did not know what they were supposed to be doing and the new managers were protecting their own position in the hierarchy.

It is interesting that about the same time TVA's Personnel Division stopped issuing its administrative codes which described division and staff functions and inter-relationships. The old system of inter-staff coordination was abolished and replaced by a hierarchy that could manage but could not define the TVA mission for employees.

These problems were to became even more serious with the first major reduction in staff that was announced just before Christmas in 1981. This was to be followed by several reductions which further eroded staff morale. Adding to the problems was a report by the Reagan transition team that faulted the TVA Board for not being able 'to set TVA on a firm course for the 1980s'.[11]

### The 1981 Reorganization

In 1981 TVA announced a new reorganization. It is interesting to note that in 1980 there was the first reference in an internal report to TVA as a 'corporation'; in earlier times TVA had always considered itself a federal agency with a clearly defined public purpose.

In announcing the new reorganization TVA officially recognized the new 'corporate' thinking. TVA would become more like a corporation than a government bureaucracy and it would have vice presidents to oversee the three major areas of TVA operation. These new positions were to be comparable to corporate vice presidents with the office managers reporting to them and handling day to day operations.[12]

Like the former re-organization this one also seemed to be based on the concept that staff reorganization would solve the agency's problems and that an clear statement of TVA mission, goals, and place in the region was not needed.

To fill this need TVA, in 1982, published *TVA: An Agency for Regional Development*. The publication noted that in the past TVA had given so much attention to the power program that TVA's economic development role had been somewhat neglected. To correct this TVA stated that:

> TVA has again assigned top priority to addressing the economic development needs of the Tennessee Valley. These needs boil down to, jobs and incomes and opportunities in a region that, despite significant progress, still lags behind the nation in most important economic categories.

> TVA is seeking to weld together its programs in electric power, natural resource development, fertilizer research, navigation, recreation, and tourism, and environmental enhancement, combined with skills in economics and regional

planning, and to apply these as a part of a single unified economic development effort in the Tennessee Valley.

This is what TVA means when it speaks of 'returning to its original purpose'.[13]

This statement seems to equate economic development with regional development. It also tells what TVA will do; it does not tell how it would do the job. It should be noted that the original TVA Board did address this issue and as a result its regional resource development program had the support of the states as well as the people of the Tennessee Valley and the South. Other than create the vice president positions this reorganization did not change significantly the nonpower resource development staff groupings. This reorganization was not to last long before other changes were made.

In 1983 Tom Ripley left TVA and the Office of Economic and Community Development was combined with the Office of Natural Resources to form the new Office of Natural Resources and Economic Development (Figure 4.1). The reason given for the merger was that TVA feared appropriations would be hard to justify for two offices and that it could operate the two as one more economically.[14]

For TVA's overall resource development program the 1981–83 reorganizations merely added to the problems created by the organizational changes of 1979. The individual professional staffs were further submerged within the new TVA bureaucracy. Without leadership at the top and with layers of vice-president managers above the overall resource development program floundered. Short term projects to solve selected community problems, and to respond to the demands of congressmen whose support was important, became the rule. Without guiding principles these individual projects had nothing to hold them together. Instead of a coordinated resource development program TVA resorted to a system of office sign-offs.

**The Merec Program, 1981–89**

In 1981 the TVA regional planning staff signed an agreement with the United States Agency for International Development (USAID) to provide planning assistance in selected less developed countries called *Managing Energy and Resource Efficient Cities* (MEREC) , the program was initiated to assist USAID projects abroad, utilizing, the TVA staff's extensive experience and competencies. The MEREC program was no doubt also attractive to TVA because it brought in external funding from another federal agency to support staff salaries at a time of decreasing support in Congress for appropriated funds. MEREC projects were first undertaken in secondary cities in the Phillipines, Thailand, and Portugal. The focus of MEREC was on integrated resource and infrastructure management, that is a systematic approach to plan development and implementation in accordance with a coordinated multi-functional strategy. Initial projects were conducted in Tacloban, Phillipines, Phuket, Thailand, and Guarda, Portugal. A number of other cities and rural areas were subsequently added, primarily in Portugal. An evaluation report prepared by TVA in late 1989 judged the MEREC program to be quite successful in assisting selected cities to undertake various projects in water supply, sewerage, housing, transportation, general planning and economic development. The report cited significant capital and operating savings in each of the participating municipalities. In the work of the MEREC program the planning staff demonstrated its ability to adapt community planning techniques to other settings quite different from that of the Tennessee Valley region. Regrettably, despite tangible successes and the popularity of TVA assistance within each country, the MEREC program was not continued in the three countries with which TVA had originally contracted nor in other countries proposed for inclusion. It was another example of a demonstration project that was successful but not repeated or more widely adopted.[15]

### The End of the Freeman Years

In the mid 1980s S. David Freeman's term as TVA's Chairman was about to end. Freeman's years were marked by notable achievements; the down-sizing of the nuclear plant construction to bring energy supply more in line with probable power service area demand; the expanded energy conservation and alternative energy program; and the reduction of pollution levels at the TVA coal fired plants all were important to bring TVA more in line with national polices.

The Freeman years also had a down-side. Although Freeman had stated that he wanted to bring TVA back to its roots he did not seem to understand that it was necessary to describe clearly the new mission of change which he had called for. Instead he tried to use reorganization as the tool for change but without a redefinition of the TVA regional mission and role these were not successful.

Finally, although Freeman was able to reduce the size of the nuclear program he did not move it in the direction of a manageable nuclear technology. He noted that the technology used by TVA was too demanding; 'There're 50,000 valves at our Brown's Ferry plant. People are going to screw up. We need nuclear technology that is less demanding, that is more inherently safe and therefore will be more economical over time'.[16] As noted above, perhaps given the heavy TVA investments in design and construction of its existing nuclear plants the necessary changes could not be made but such changes could have contributed much to national energy policy and hopefully, long-term, could reduce TVA nuclear construction costs. Freeman left TVA on May 18, 1984 noting 'I've given it my best shot'.

Charles 'Chili' H. Dean, Jr., the General Manager of the Knoxville Utilities Board, had been appointed to the Board by President Reagan in 1981 to fill out the term of Robert N. Clement who had resigned. At the same time he had appointed Dean Chairman of the TVA Board. Reagan also appointed John B. Waters, Jr., a lawyer and close associate of Howard Baker, to the TVA Board.

As in the case of Freeman, Dean inherited the two problems that had plagued TVA for more than a decade: the safe design, construction, and maintenance of the nuclear plants and the need for a redefinition of the TVA regional development mission.

## The Nuclear Plant Problems Continue

Prior to the departure of Freeman the TVA Board had approved the merger of the Office of Power and the Office of Engineering Design and Construction. It became the responsibility of Dean to carry out the merger which was scheduled for June 1, 1984. The Office of Engineering Design and Construction had opposed the cancellation of so many of the nuclear projects and it is hard to judge how this position affected the proposal for merger with the Office of Power. The merger did consolidate the overall dominance of the Office of Power and reduced further the number of independent views presented to the Board. Perhaps with the decline in the size and variety of the TVA construction program the merger was a logical step in trying to deal with the problems facing the power program and particularly the problems associated with nuclear plant safety, operation, and maintenance.

For the professional staff of the Office of Engineering and Construction this must have been a difficult time. TVA had always prided itself on the excellence of its engineering personnel and the huge construction programs which this staff had designed and managed. It certainly represented a major change in the way TVA was to consider future construction projects. In addition the reorganization served to further submerge another important professional staff within the TVA organization in much the same way that the 1979 reorganization had done to the community and economic development staffs.

The merger may have been a part of TVA's concern that the nuclear plant problems had not been resolved. As a part of the merger the Office of Power proposed reducing the central office staff and placing more responsibility and resources at each nuclear plant with the site director having responsibility for activities at the plant. This resulted in 120 to 150 nuclear power employees from Chattanooga and 150 engineering design employees from Knoxville being transferred to Browns Ferry, Sequoyah, and Watts Bar nuclear plants.[17]

These reorganization efforts did not solve the problems at the nuclear plants particularly those at the Watts Bar Plant. The Nuclear Regulatory Commission raised questions of the safety of the plants as well as TVA's ability to problems of quality control and management of the plants. The

**Figure 4.5   Earle Draper, planner of Norris, receiving an award from TVA Chairman, Charles Dean (r.) at the 50th Anniversary Celebration of Norris, Tennessee, 1983**

*Source:* D.A. Johnson.

result was that NRC repeatedly postponed the licensing of TVA nuclear plants.

In addition there were charges of discrimination against employees who raised questions about the safety of the TVA plants. The Nuclear Regulatory Commission was becoming critical of the whole TVA nuclear program. In fact, a Department of Labor investigation in 1985 cited the NRC's Atlanta staff as saying that the TVA Board and General Manager were neither a positive nor negative force in the control and management of the nuclear program. They were simply 'irrelevant'. They added 'In the absence of control from above, TVA's separate divisions degenerated into feudal fiefdoms which were controlled by bureaucratic princes competing for power'.[18]

Major changes where made in January 1986. First, TVA created the Office of Inspector General and named Norman A. Zigrossi, formerly in charge of the FBI field office in Washington, to that post. Second, it put Steven White in charge of TVA's nuclear power program. White, a retired four-star admiral with a 33-year naval career, was brought in under a $29,000 per month contract. In addition, Stone and Webster, a national engineering consulting firm with which White was associated, was given an $8 million contract to provide managers for key positions.

The presence of White on the scene seemed to give a new confidence that TVA would be able to solve the problems of its nuclear program. White, however, insisted this was a long term job which could not be accomplished overnight. But for the moment the nuclear program seemed under control. Then, in early 1986, another event occurred that was unsettling for TVA. Unannounced on February 13, Richard Freeman resigned from the TVA Board. This left TVA once again with a two man board, now consisting of Charles Dean and John Waters. But the problems with the nuclear program continued. Less than two years after his appointment as TVA nuclear chief Steven White faced a problem with the Justice Department over a NRC charge that he had lied under oath. Though the charges were dismissed, they had cast new doubt on TVA's nuclear program. Moreover, TVA faced continued NRC concerns over the safety of its nuclear plants. In early November 1987 NRC noted 64 problems at the Watts Bar Nuclear Plant including some major structural deficiencies that would require solution at the plant before it could be licensed for restarting. Steven White had been right in stating that correcting TVA's nuclear plant problems was a long term job.

## The Search Continues for a Regional Development Program

At the end of the 1980s the nuclear power problems were still having repercussions on the overall TVA program, a situation that was recognized by General Manager William F. Willis. He outlined his concerns at a 1985 conference of TVA managers on the future direction of TVA.[19] Willis opened the conference by describing what he saw as the lessons of the past years. These included prevention of future failures such as had occurred with the nuclear program so as to sustain pubic confidence in TVA as an institution; maintain TVA credibility as a vital agency; find ways to inspire managers and above all create an atmosphere of trust between employees and supervisors at all levels; work out new relationships with the valley states; and active staff help to the TVA Board in solving the problems of TVA relationships with the Tennessee Valley states. A central point to all problems seemed to be the erosion of TVA independence that 'impedes us from being as responsible, in being able to act as quickly and effectively as you can in some corporations'.[20]

Willis followed by describing the TVA mission in terms of seven objectives and a five point action agenda. The objectives which Willis used to define the TVA mission were (1) maintain competitive power rates and regain TVA's role as a yardstick of excellence for the utility industry; (2) assure for the Tennessee Valley Region a reliable and ample power supply; (3) excel in the management and operation of Tennessee river and reservoir system as an important element in the continued economic development of the valley region; (4) establish a role in economic development that is unique to TVA and is so recognized by TVA's regional partners; (5) effectively manage the National Fertilizer Development Center at Muscle Shoals; (6) make TVA a national demonstration agency which could help to solve major problems facing the nation; and (7) plan and operate in a way that protects the natural environment.

Willis's action agenda mirrored both his lessons from the past and his recommended objectives: (1) get the nuclear program back on track – a problem for everyone directly connected to TVA; (2) demonstrate TVA's competence regularly, systematically, and credibly; (3) inspire and excite managers and employees to work effectively together; (4) form new partnerships with the states; (5) define and communicate TVA's national mission and yardstick role.

Much the same point of view was outlined in TVA's *Multi-year Plan for 1987–1992*. This plan used the seven objectives noted above and added two more: create and maintain a work place that encourages all employees to attain their fullest potential and set the standard for federal agencies in efficiency, productivity, and financial management.

It is interesting that all of the objectives outlined by Willis are essentially operational. Moreover it is questionable if they constitute a restatement of the TVA mission; in fact these statements merely call for a search for a new mission. In many ways the Willis perspective illustrates the problem TVA faced as a regional agency which had completed its assigned specific duties to improve the Tennessee River for navigation, flood control, and power and to integrate these specific elements into a broad regional development program. Once these tasks had been completed TVA needed to redefine its mission in new specific terms and in terms that were unique to TVA as a regional agency – a task which it seemed unable or unwilling to do. It is true that this would have involved opening the TVA Act to major changes which TVA had always been reluctant to do for fear it could not control the outcome.

In addition, it seems clear to the authors that even the power program needed a new mission because the original yardstick mission had been accomplished. Given the nuclear mishaps at TVA's own Browns Ferry in 1975, at Three Mile Island in Pennsylvania in 1979, and at Russia's Chernobyl plant in 1986 which raised so many national concerns, should the TVA power mission merely have been to make its own nuclear plants work? Or was there a larger national concern relating to energy resources which TVA might have addressed if it wanted to be as Willis had urged 'a national demonstration agency'. If TVA wanted to embrace nuclear as it clearly had done over the last twenty years why did it not seek to demonstrate for the nation a system of safe, easily built and maintained nuclear plants? Within this mission most of the Willis objectives would have been applicable. While Willis supported this larger view of the power program, those in charge of power were so caught up in building and operating the power plants, they did not understand the importance of relating the power program to a national demonstration mission.

As already noted maybe this was expecting too much of TVA. Why try to secure approval of such a new mission when the very life of the agency might be at stake? But TVA originally was not an ordinary agency and it could have pushed ahead in spite of the risk. Because it did not act TVA became just another government bureaucracy protecting itself and operating as another public utility. In this instance and into the future it was to become incapable of not only defining a new mission but of solving such a basic organizational problem as the maintenance of employee morale. To illustrate, a 1986 report noted that sixty-four percent of the employees who voluntarily resigned from TVA did so, in part, because of dissatisfaction with upper management and agency policy.[21] But this had been a problem since David Freeman was Chairman and had reorganized the agency without a clear understanding of mission or purpose. It is interesting to note that within TVA the discussion of the problem was always focused on the attitude of managers and never on the programs the managers were to implement or how programs were related to the so-called 'change' within TVA.

Another example of a TVA financially supported regional development program which lacked focus was the rural economic alternatives program of the East Tennessee Community Design Center, a private, non-profit agency. This program, operated by the Design Center, was funded by a startup grant from TVA and was to cover a 45-county area in East Tennessee, Southwest Virginia, and Western North Carolina: the combined service area of the East Tennessee Design Center and the Appalachian District Office of TVA.

The Design Center noted in its Summer 1987 newsletter that the program had these five characteristics: an Appropriate Technology attitude toward rural Appalachian economic development; employment of a Rural Development Coordinator to broker technical assistance to the region; the support and nourishment of grassroots community developers; a strong community economic development training program; and a revolving fund to provide financial resources at critical points in the business development process.

Looking at the demographic and economic changes which had taken place in the region over the past 50 years, what did TVA mean by rural economic alternatives when it funded this program? It is difficult to see how this program, which lacked a strong urban component to relate program activities to regional trends, could do much more than support the status quo by keeping people in situations where even the children could not escape. John Kenneth Galbraith in his book *A Journey Through Economic Time* called the pockets of poverty caught in the isolation of mountain areas as 'prisoners of both past and present' because they lacked access to wider economic opportunity and the needed education 'that are the hard substance of economic progress.'[22] In addition, this program, without the active involvement of state community and economic development agencies, had little chance of success. The program seemed to be premised on past agrarian views of the South and the Tennessee Valley Region. Again this effort reflected the inability of TVA to articulate a regional development mission for itself which met the needs of a region that was already urbanized and mirrored national demographic and economic patterns and which in many ways reflected the economy and life style of the nation. As a part of its regional development program TVA also started a reassessment of the entire TVA reservoir system. While this assessment did result in changes in reservoir operations, there did not appear to be an overall plan for relating reservoir shoreline land uses to current valley needs.

1986 and 1987 were to be continuing turbulent years for TVA. First, Congress became increasingly critical of the seemingly inability of TVA to bring the nuclear program under control and in full operation. Then there were budget cuts which threatened TVA's non-power program. A drought in the valley strained the river control system and evoked charges that the TVA river control system was overpowering the rivers and bringing about a decline in water quality. Finally, over protests of many groups the TVA Board raised power rates by 7.8 percent.

All these problems brought into question the credibility of TVA and threatened its very existence. A report of the Southern States Energy Board, an advisory group to the Southern Governors Association, called for expanding the TVA Board, the lifting of pay caps on all TVA salaries, and the placing of the 'utility's' day-to day operations under the control of a chief executive officer hired by the TVA Board. The report claimed that the present structure of the TVA Board caused 'group think' and lead to the failure of the TVA nuclear power program.[23] Senator Albert Gore came to the support of TVA and, as if anticipating the next move by TVA critics, proclaimed TVA would never be for sale.[24]

Facing some of the most difficult times in its history TVA started looking at options for a TVA revival. In reviewing the record of those times the frantic mood of the TVA staff stands out. Again, instead of first attempting to clarify a new TVA mission as a basis for action it turned to 'reorganization options'. The Office of Policy, Planning, and Budget, headed by John Stewart, had been asked to look into potential options for changes in TVA's organization and structure. Stewart seemed to believe that the core question was 'can you function as a business or can't you'.[25] This in turn seemed to translate largely into how TVA could bypass federal pay restrictions so that its top power managers could be paid higher salaries. Stewart's 1987 final report entitled *Options for the Structure and Organization of TVA* examined seven organizational and six legislative options. Certainly the most drastic of the options suggested was to create a TVA Holding Company with two independent subsidiaries; a self financed Tennessee Valley Power Corporation which might evolve into the regional public ownership of TVA's power system by power distributors; and a separate Tennessee Valley Development Corporation financed by congressional appropriations.

But another change was on the horizon that was to cancel further consideration of these options. In late September 1987 President Reagan announced his intention to nominate Marvin T. Runyon, President of the Nissan Manufacturing Corporation, USA to be Chairman of the TVA Board. This meant that Charles Dean would turn over the chairmanship of the TVA Board to Runyon and that for the first time in two years TVA would have a full three-member board.

### The Runyon Years Begin

Marvin Runyon was unique as a TVA Board member and chairman. He was to become the first board member who had not had at least some experience in government. At the time he took office Runyon was 63 and had spent 44 years in the automobile industry; 37 of these years with Ford Motor Company where he had risen to the position of vice president for body and assembly operations. In 1980 he moved to Nissan to set up its operations in the United States. He had selected the Tennessee site at Smyrna and had put the $161 million plant into production in 1981.

Runyon promoted a participatory theory of management – one that would bring all involved people into the process. He used this concept at Nissan where reportedly, it was not unusual to see him walking through the assembly plant dressed in the blue regulation Nissan workclothes talking to the on-line employees about the problems they were experiencing. Representative Cooper, the head of a group of congressional members from the Tennessee Valley states, believed that Runyon's reputation as an effective administrator would reduce the demand for an expanded TVA Board. Runyons's appointment did in fact result in a lessening of demand for changes in the TVA Board structure.

Runyon was sworn in as Chairman of the TVA Board on January 25, 1988. Almost immediately the Board changed the way it did business. At its first meeting with Runyon as chairman, the board decided to divide responsibilities among the three directors aimed at 'improving efficiency in the management of TVA programs'. Under this arrangement Runyon would provide guidance on all administrative matters which included finance, budget and planning, employee relations, governmental and public affairs, corporate services, and TVA's Inspector General and General Counsel; Charles Dean on power matters; and John Waters on matters affecting those TVA programs financed with congressional appropriations. The board members would work together 'as a team at the top, sharing their knowledge as they deliberate the overall policies that guide the agency'.[26]

This Board organization was termed an 'innovation' but it was much like the arrangement originally used by the first TVA Board in the 1930s, but later abandoned when it proved to be unworkable. The difference was that under the 1988 arrangement the Chairman, Marvin Runyon, had the support and confidence of the other two board members.

Although General Manager Bill Willis was to continue to make day-to-day decisions on the implementation of Board policies it is interesting that the Board also formed a top management team which was created to get 'everyone working together as a team' and which would meet weekly 'to plan and manage TVA operations'. This was followed by Runyon's appointment of six panels to study TVA by focusing separately on power organization, power needs, non-power appropriated programs, legislative initiatives, business management, and staff pay.

Runyon had clearly put the TVA staff in a positive mood. Employees were enthusiastic about the pledge of the new chairman to listen to them. But the Runyon approach also raised some questions. First, instead of a regional development program which would include power as well as all TVA activities, the programs funded by Congressional appropriations were placed in a separate study category which made it more difficult to define clearly the mission of TVA as a regional development agency and more importantly, its place in the region. Second, in spite of Runyon's commitment to a participatory style of management 'where people at the top give up some control of the process and push responsibility down to the employees' it is interesting that only top managers were included in the study panels. Employees were urged to provide suggestions and comments to the individual study panels.

The Runyon 'honeymoon' did not last long. Less than a month after taking office there were objections to the new TVA Board policy on how the board handled public comment. Groups generally concerned with TVA actions concerning environmental matters felt that there should be more citizen involvement in TVA's announced 'participatory management decision-making'. The objection came because of a new board policy that no longer permitted citizens to question board members during a board meeting. Citizens could comment on agenda items and ask questions after the session. Runyon defended the practice by saying it was normal business practice.[27] As time passed it became clear that Runyon did not see TVA as a government

**Figure 4.6   Marvin T. Runyon, Chairman of the TVA Board, 1987–92**

*Source*: TVA file photo.

agency but rather as an independent public utility much like Duke or Alabama power companies, a business that had customers and had to be competitive to hold these customers.

In March of 1988 TVA received welcome news that the Nuclear Regulatory Commission had approved the restart of Sequoyah Nuclear Plant Unit #2. By this time, however, Runyon had become concerned with TVA's $18 billion debt – a debt which could only rise in the future given the continued problems in the nuclear program. This concern was certainly a major reason for Runyon to seek ways to reduce costs and maximize income. It is strange, however, that Runyon, given his long experience in private business did not attempt to limit the growth of the debt. This was to have to wait for the appointment of Craven Crowell to the board. In September 1994 Crowell put a self-imposed limit on the TVA debt.

By late April, 1988, Runyon working with the Management Committee, announced three objectives which he defined as the TVA mission: (1) to operate a more competitive power system by holding the TVA's business firm power rates constant for three years – meaning no rate increases; (2) to provide better services to the people of the region from federally funded programs by reducing overhead costs; and (3) to establish a more business-like organization and to focus on efficiency, productivity, and accountability. Runyon then told employees these steps would require TVA to reduce its work force between April and the end of the fiscal year on September 30, 1988. He again asked for employees to become involved in the process with the statement 'we'll be communicating decisions and a timetable for their implementation as soon as possible'.[28]

In view of the fact that Runyon in talking to employees had posed the fundamental questions 'What are you doing' and 'Why are you doing it', it is noteworthy that Runyon's announced objectives did not include a substantive statement of mission which might have helped employees answer those questions. According to the media several of the top TVA managers echoed Runyon's remarks saying these are the questions 'we should have been asking ourselves'.[29] It would seem that the job of the managers should have been to clarify and articulate the mission for employees. It is not surprising that employee morale once again declined and, under stress as to their future, their efficiency and productivity declined.

As had been the answer in the past, TVA once again turned to reorganization as an answer to its problems. This new reorganization again promoted a structure resembling a private corporation; it established three new vice-presidencies reporting directly to the Board rather than to Bill Willis, whose title had been changed from General Manager to Chief Operating Officer. In the view of University of Tennessee Professors Max Wortman and David Welborn the organizational changes were regarded as giving Runyon greater power over budget and personnel matters and over the character of TVA[30] Runyon in effect, became both Chairman of the Board and Chief Operating Officer of TVA.

A big personnel announcement came on June 29, 1988 when Runyon announced the layoff of 7,500 (22 percent) of TVA's employees with assurances that employees would be told of their status immediately. Runyon was later forced to admit he made a mistake in making that commitment when it was weeks before some employees were to learn of their status. Employee morale was not to recover during Runyon's tenure on the TVA Board.

The reorganization also served to further obscure TVA's role as a regional development agency. With the exception of power all other TVA line activities were included in one group entitled 'Resource Development' which had three divisions: Business Operations, National Fertilizer Development Center, and River Basin Operations. How this type of organization meshed with goals described in the report, *TVA Corporate Environmental Agenda – 1990*, is not clear. These goals were described as follows in this statement:

*LAND RESOURCES* – practice and promote land use planning and management that enhances natural and economic values.

*WATER RESOURCES* – maintain and enhance Valley water resources for sustained human use and enjoyment, economic development, and biological diversity.

*AIR RESOURCES* – sustain a level of air quality that is healthy and supports economic development in the Valley.

*COMPLIANCE* – plan and conduct TVA's own business in a way that ensures compliance with environmental regulations and agency policy.

*MATERIALS AND WASTE MANAGEMENT* – practice and promote materials and waste management which emphasizes an efficient life-cycle.

Except for TVA's own operations, the statement did not indicate how it was to work on these goals in areas where it had no control or why these goals were unique to TVA as a regional development agency.

These efforts by Runyon to restore morale and a sense of mission to the TVA staff raises interesting questions. First, while Runyon's participatory management style probably worked well where goals and mission were singular and clear, as in the production of automobiles, it did not seem to work where the mission was complex and not readily understood by the staff working on a broad regional mission which involves multiple programs. The evidence seems to be that his management style did not cure low employee morale or make the overall regional program more effective and productive. Runyon also brought in top managers whose training and experience was in private business without any knowledge of regional development. This, plus ever-changing policy and administrative leadership, resulted in employees being subjected to the whims of the administrators without the leadership the employees had the right to expect from the program managers. The result was that employees, once again, did not know what was expected of them or what to do. It certainly did not produce efficient and productive employees.

A second and even more intriguing question relates to the power program itself. The TVA Act outlined certain social objectives for the power program which were much broader than the Runyon requirement of being competitive. Under the Act TVA was to establish a yardstick for the power industry and was to give preference to domestic use of power so as to improve life generally in homes and farms of the Valley Region by making available abundant and inexpensive power. These objectives had been achieved long ago. Moeover, the Tennessee Valley had moved into the mainstream of the national economy. Merely being competitive did not outline a new present day social objective for a public power system such as the one noted previously; the demonstration for the nation that a safe and easily maintained nuclear power system was possible. In reviewing the record it is not clear if the TVA Board and top management were even concerned with national policy. For example, a secondary result of the cost cutting and job eliminating efforts was the cancellation in 1988 and 1989 of most of its energy conservation program.

Certainly the question has to be asked, would a more logical step in trying to solve TVA's power and regional development problems have been to accept TVA as a publicly owned utility system rather than a regional development agency. TVA could have supported the kind of economic development program operated by most private power companies. This would have permitted scaled-back reservoir operations with some of these operational responsibilities going to other federal agencies or to state and local governments. Or TVA could have continued such operations with congressional appropriations covering those expenses not directly chargeable to the power program.

There were other personnel changes which seemed to reflect Runyon's views of TVA as a private corporation rather than a pubic agency. There were disagreements with the progress Steven White was making in dealing with the nuclear power problems. Oliver Kingsley was subsequently brought in as a TVA employee to replace White as head of the power program. A new effort was made to find ways to increase compensation for TVA employees. An executive bonus system was created to circumvent the cap on federal pay. This seemed to ease a limiting factor which TVA had long considered a handicap to hiring highly trained and experienced people in the nuclear field.

Runyon also brought in William Malec as Chief Financial Officer. Malec started a program to restructure the TVA debt and to find ways to control costs and reduce expenditures. Interest on the debt accounted for 34 cents of every dollar TVA spent – a total of $1.8 billion annually. Most of the TVA debt had been financed by the Federal Financing Bank (FFB). FFB would not permit TVA to refinance its high interest debt until the callable date on the bonds. TVA, on Malec's recommendation decided to take advantage of the lower interest rates on the open market. In mid-October 1989 TVA issued $4 billion

**Figure 4.7    William F. Malec, Chief Financial Officer, 1990–1994. To reduce TVA interest payments on its debt Malec refinanced some of the high interest bonds for lower rate bonds purchased on the open market**

*Source*: TVA Staff Photo.

in bonds, a move that resulted in interest savings of $75 million annually.

In July, 1990 Malec announced that TVA had saved $168 million by refinancing its debt in the open market, lowering its average interest rate on the debt from 10 percent in April 1989 to 8.5 percent in July 1990. While important in reducing interest charges, with the debt rising at a rapid rate these actions would appear to be only stop gap measures. Although it is true that TVA did make major savings in its interest payments on existing debt, with further borrowings the debt had risen from $18 billion in 1988 to $21 billion in 1990 with the result that TVA's interest payments actually increased.

Runyon also tried to focus TVA management and staff on its overall mission. In April 1990 Runyon announced the results of board meetings with over 200 managers on a strategic plan to shape TVA in the 1990s. The plan included five areas for TVA action: (1) environmental leadership that goes beyond laws and regulations: (2) stimulating economic growth with attractive power rates as the best incentive TVA can provide for new growth; (3) excellence in consumers service through a consumer first philosophy; (4) maintaining a quality workforce and workplace by stabilizing the TVA workforce, expanding employee involvement, and continuing to improve productivity; and (5) excellence in nuclear power operations by operating TVA plants in a safe, efficient, and quality manner.[31]

As usual TVA turned to a major reorganization to carry out this strategic plan. In January 1991 the TVA Board announced a plan to divide the agency into three broad operating groups: generating, resource, and consumer. The plan did away with the Chief Operating Officer position held by Bill Willis who in turn was made President of the Resource Group. Oliver Kingsley headed the Generating Group, and Mary Sharpe Hayes headed the Consumer Group. An executive committee composed of Kingsley, Sharpe Hayes, financial officer Malec, general counsel Ed Christenbury, with Bill Willis as chairman reported directly to the board with responsibility to shape TVA's long-term business strategies, recommend major program initiatives, and guide day-to-day operational decisions.

With employee morale already low TVA tried to assure employees that the reorganization would not cause immediate layoffs but indicated that some jobs could be 'changed, eliminated, or moved'. In spite of these assurances TVA announced plans to disband its construction work force by May 1991 and contract to outside firms 'non-routine' work such as plant completion, major retrofit, and specialized maintenance and refueling. The move affected 6,300 construction workers and some 300 administrative employees. As a result, although Runyon could point to reduction in the number of permanent TVA employees, the total number of people working on TVA projects – TVA employees plus contract workers – did not decline significantly.

By June 1991 a U. S. General Accounting Office report stated that TVA management and labor relations had gotten so bad that it recommended that Runyon attempt to mediate some of the problems. If not done in six months Congress should act to make TVA comply with federal labor relation laws from which its is now exempt.[32] TVA justified its actions by noting expected savings of $105 million annually by the awards to private contractors.[33] But not only construction workers were affected by this policy. In August 1991 TVA announced that it was hiring the private architectural firm of Earl Swensson of Nashville to assist the agency in architecture and design. The firm was to guide TVA architects. When one looks back over the remarkable achievements and awards of the TVA Architectural Staff for design of TVA dams and power plants it is difficult to understand what the board and presumably the executive management committee was trying to achieve with this move.

TVA did set up an employee transition program for the permanent employees displaced by disbanding the construction workforce. But many employees did not believe the program was adequate to meet the problems when so many employees were losing their jobs. TVA claimed the program was working well and it did for a small number of employees.

If the operating personnel were having difficulties, TVA top management was doing well. In 1990 TVA paid a total of $1.3 million in bonuses to 68 of its executive level managers; doubling the 1989 total. In 1991 it paid $1.88 million to 88 top officials.[34] With the major nuclear problems still unresolved and employee moral at its lowest level in years it is difficult to understand the justification for the bonuses.

### The Regional Development Program – 1990–1991

Some of the actions taken by TVA during this period also make it difficult to understand the direction and thrust of the TVA regional development program. *Inside TVA* for January 15, 1991 and January 14, 1992 reviewed resource development activities for 1990 and 1991. These reviews provide information on the scope of the program for those two years.

In 1990 Board Member John Waters became interested in the development along the main river system of TVA lakes and started what was called 'the River Heritage Program'. A specially prepared TVA towboat and barge carried Waters and TVA staff members on a highly publicized tour of the 650 miles of main river lakes; stopping at communities and cities along their shores. They met with over 1,000 people to discuss developmental problems relating to the communities and the river. TVA also started work on a Lake Improvement Plan which evaluated how TVA managed the Tennessee River and lake system and the changes that Valley citizens thought were needed.[35]

At the urging of Senators Jim Sasser and Al Gore, Jr. TVA began looking into the possibility of burning combustible trash for the production of electric power. Here, too, TVA looked only at internal TVA costs and problems. Both Sasser and Gore noted that waste disposal was a regional as well as a national problem and suggested the need for new approaches as a demonstration. If TVA had looked at the nature of such problems as waste disposal and

**Figure 4.8**    John B. Waters, lawyer and political advisor to Sen.
Howard Baker, appointed to TVA Board by President
Reagan in 1984, appointed TVA Chairman by President
George H.W. Bush in 1992; served until 1993. Waters'
special interest was a program he termed 'River Heritage'.
He used a TVA towed barge to visit river communities to
discuss reservoir and operational improvements

*Source*: TVA Staff Photo.

protection of water resources it should have understood that new institutions were needed to deal with the problems. But TVA had lost its broad-ranged technical staff and its managers did not understand the nature of regional development and shared powers. As a result TVA lost yet another opportunity to demonstrate its ability to serve as a regional development agency.

Additional activities in 1990 included awarding $30 million in contracts to minority enterprises and restructuring the fertilizer development center at Muscle Shoals to become the National Fertilizer and Environmental Research Center. The common thread to all these activities seemed to be that all were related to TVA operations rather than to a broad regional development program. The activities of the 'Resource Group' in 1991 followed much the same pattern. The activities for that year included: the approval of A TVA Lake Improvement Plan with new operating priorities that kept lakes at their high summer levels longer so as to benefit recreation and tourism; improvement of lake public use facilities to accommodate increasing recreation use; a new working partnership with the Appalachian Regional Commission to promote cooperation on valley economic development problems; demonstrations of environmental safeguards at sites of fertilizer manufacturers and dealers throughout the country; construction of a artificial wetland research and development facility at the National Fertilizer and Environmental research Center; funds for waste research projects in nine states; moving the TVA and Small Business Administration mobile assistance center throughout the Tennessee Valley to promote business development weeks in small towns; and the design and development by TVA of the first two-way interactive video networks to link schools and communities to community learning centers. It was during this period that the TVA Flood Damage Prevention Program which had become a national demonstration for treating flood prone areas as a part of the state and local land use planning problem began to revert to the pre-1950 practice of considering only engineering solutions to local flood problems.

One activity deserves special note. In the 1980s the Tennessee Valley was not organized to promote technology-based economic development. Recognizing the research and development resources in the Knoxville-Oak Ridge area, TVA joined with the University of Tennessee, the Oak Ridge National Laboratory, and the Department of Energy/Oak Ridge operation to set up a series of institutions to help promote the transfer of technology to new and existing commercial operations. The Tennessee Technology Foundation was set up in 1982 and staffed with personnel on loan from TVA. Over the next six years this group helped 140 new, expanded, or relocated business use the technology developed in the area to create over 2,000 jobs. A Consortium of Research Institutions was formed in 1983 to link the research and development work in the Knoxville-Oak Ridge area. Out of this came TVA's technology brokering program – a program to match valley technology firms with Federal funding agencies. In 1988–1989 over $3 million in Federal R&D funds came to TVA which used the funds to encourage valley industry to expand their technical expertise to help them compete with firms from other regions which already had well established firms with such expertise. Although this was a sound program initially it probably was not monitored sufficiently to prevent abuse.

During August and September 1991 the Defense Department transferred some $96 million to TVA to contract for modernizing ships and designing a machine gun range because the Defense Department could not award these contracts before the end of the fiscal year when such funds would have to be returned to the U. S. Treasury. TVA could speed up the process by not having to seek competitive bids for defense work. An activity involving the exchange of this amount of money would certainly have to be approved by the Board as well as top TVA management and the TVA legal staff. Willis as President of the Resource Group and Chairman of the 'executive committee' had an easy avenue to bring this to the attention of the TVA Board and top TVA management for discussion and agreement on its policy implications. High level staff thought TVA could use these funds to place contracts with Tennessee Valley companies and thus encourage and expand high technology industry in the area as well as receive a management fee of up to 10 percent to recover TVA costs. In many instances, however, the Defense Department dictated the firms to receive the contracts and as a result much of the money went to firms outside the region.

Later a Senate subcommittee was to claim that TVA as well as other agencies broke the law in handling these contracts for the Pentagon. The Defense Department Inspector General made a finding that defense officials acted improperly in shifting responsibility for contract funding to TVA. Media reports note that Willis asked TVA's Inspector General to review the TVA actions and according to such reports he found that TVA handling of the contracts often did not benefit the region. Finally in 1993 amendments were made to existing laws to stop the Department of Defense off-loading contracts to other agencies. In mid-July, 1994 TVA repaid to the U.S. Treasury the $6.8 million fee it collected for managing defense contracts for the Pentagon.[36]

Here, too, it is difficult to see a pattern in these separate projects that could be considered a regional development program. Instead most of the activities seem to be related to TVA's own operations or to individual projects rather than a group of complementary programs to solve specified regional problems. In addition there was no indication of the region to which many of the projects were related. For economic development the river basin is an inappropriate region, as Levin had noted in his 1968 critique of the TVA regional development program. (Levin, 1968).

## New Managers and New Organization for the Resource Group

In early April 1992 the Resource Group Staff prepared a statement of the group's initiatives for the 1990s which had as its main thrust 'to develop and implement environmentally sustainable solutions in the areas of natural resources, energy, and economic development that improve the quality of life in the Tennessee Valley and the nation'. Willis indicated six areas of major concern: clean water; jobs creation/increased income; workforce and environmental education; waste management; model resource stewardship; and excellence in agency environmental and technical support.[37]

This report was hardly off the press before the Resource Group Staff learned it was to have a change in management. On April 16, 1992 William Willis was removed as President of the Resource Group and given the job as President of a newly formed Board Advisory Group which was actually the reconstituted executive committee established initially by Runyon. Nine months later, on February 7, 1993 Willis retired leaving TVA without any managers with an understanding of TVA's role as a regional development agency.

Runyon named Norman Zigrossi, TVA's Inspector General, to take Willis' position as President of the Resource Group. Given the fact that this group had gone through a reorganization only 15 months earlier it is not surprising that employee morale reached a new low.

But this was not the only change to affect the TVA staff. On May 4, 1992 Runyon announced that effective July 2 he was leaving TVA to become Postmaster General. Although Runyon noted in a letter to employees that 'during the past four years the men and women of TVA have created a new chapter in TVA history, one that has brought TVA into a new competitive era', the question remains, had the basic TVA problems been solved? Although TVA electric power rates had remained constant the problems of the nuclear plants had not been solved; employee morale was still low; a workable regional development mission had not been articulated; and the problems of a high and growing debt remained.

After his resignation from TVA Runyon wrote a letter, published in the July 5, 1992 issue of the *Knoxville News-Sentinel*, in which he defended the TVA nuclear program and then noted 'I have expressed TVA's support for one-step nuclear plant licensing, standardized designs, and greater international cooperation in sharing innovative construction approaches and technologies. Advances in each of these areas can help nuclear power provide an even larger share of electricity that America and the world will need in the 21st century.'

This was not much different from the position taken by S. David Freeman when he left TVA. The question is why neither Freeman nor Runyon had been able to redirect the nuclear program while they were on the TVA Board. Runyon left a TVA that was more a private corporation than a public agency with a well-defined public mission such as he had outlined in his letter to the *News-Sentinel*. Moreover, under Runyon the driving force for TVA employees became a consumer-driven commitment to cost cutting so as to be competitive and efficient. Merely being competitive did not provide a new public purpose for a public agency which had completed its original mission and which was now operating in a region vastly changed from the Tennessee Valley of the 1930s.

On June 29, 1992 President George H. W. Bush named John Waters Chairman of the TVA Board. But Bush did not appoint a replacement for Runyon. Thus the TVA, once again, was to operate under a two-man board.

## The 1992 Strategic Plan

Chairman Waters continued the directions outlined by Runyon during his tenure as Board Chairman. In 1992 the Corporate Planning Committee headed by Consumer Group President Mary Sharpe Hayes released its Strategic Plan outlining the broad direction for TVA's future.

This plan focused on three strategic planning themes – competitiveness, customer focus, and innovation. The plan outlined what the committee saw as the major TVA purpose: 'to serve our region and nation by leading the way to quality economic growth based on competitive energy supply, effective management of the Tennessee River, demonstrated environmental excellence and innovative partnerships for community development'.

The plan then spelled out the TVA mission applied to five functional areas:

*WORKFORCE:* to create an environment in which high performance is rewarded, teamwork is expected, diversity is embraced, risk-taking and innovation are encouraged, and continuous improvement is a part of everyday life.

*ENERGY:* to be America's first choice of electric power by the year 2000.

*ENVIRONMENT:* to achieve environmental excellence and leadership by the year 2000.

*RIVER:* to provide leadership to make the Tennessee River the cleanest and most productive commercial river system in the United States by the year 2000.

*COMMUNITY PARTNERSHIPS:* to help valley communities demonstrate sustained economic progress 10 percent faster than the national indexes by the year 2000.[38]

There was one notable change in TVA policy. In 1988 and 1989 Runyon canceled most of TVA's conservation program as a part of his cost-cutting effort. Less than a month after Runyon resigned TVA hired staff to again promote energy conservation.

In comparing the 1992 Strategic Plan to that devised by the original TVA Board of the 1930s it should be noted that the original plan was based on specific and defined problems facing the Tennessee Valley Region and the South whereas the 1992 plan related basically to TVA operations or internal problems, with the exception of 'innovative partnerships for community development'. The purpose of these relationships or the problems to be addressed by the relationships were not outlined in the plan. The 1992 TVA Annual Report supports these conclusions. The major accomplishments under 'competitiveness' and 'customer focus' all related directly to TVA operations. One of the most significant actions taken by TVA was to reduce electric power costs to business and industry; an action which Chairman Waters claimed had encouraged 2,200 industries to locate in the Tennessee valley region creating 125,000 jobs. Under 'innovation' two of the major accomplishments listed related to employee relations and the remaining three related to wetland research, burning shredded tires at the Allen Fossil Plant to test a potential solution to the problem of scrap tires, and an effort by TVA foresters to find new ways to test the health of the nation's forests.[39]

The TVA Board and top management made a major effort to inform TVA employees of the strategic plan and to address the concerns of employees over workplace problems. But the problems of employee morale would not go away. The results of an overall employee survey indicated high percentages of dissatisfaction in almost all categories. The poor ratings were particularly high in questions relating to employee welfare, communication, change and reorganization, leadership and career development. These surveys cover all TVA employees. No figures were available on individual groups but after talking to many employees we believe the dissatisfaction ratings with TVA must have been much higher for the Resource Group. This conclusion should not be surprising given the many reorganizations and changes in management to which this group had been subjected over more than fifteen years. Moreover the group was to face still more changes almost immediately.

## The 1992 Resource Group Reorganization

As noted above, Chairman Marvin Runyon, before he resigned to become Postmaster General, had named Norman Zigrossi, the TVA Inspector General to head the Resource Group. In November, 1992, a little more than six months after his appointment, Zigrossi announced major management and organization changes for the group. Changes of the magnitude proposed certainly had to have the approval of the TVA Board and other top TVA managers. In spite of the TVA Board's stated commitment to 'employees first', as had happened in most past changes in management and organization, the decisions for change were made by the Board and top management without much regard for employee views or welfare.

In announcing the reorganization, Zigrossi noted that the Resource Group was refocusing its efforts to reduce overhead and to increase the amount of money going to local communities. With this reorganization he saw the Resource Group as being better 'prepared to fulfill its mission and help TVA meet the challenges of the 21st century – providing jobs while protecting the environment'. He did not describe the mission or the challenges TVA would have to meet in the future. Like so many of the top TVA managers of the past 15 years Zigrossi dodged the issue of mission by calling for change although he did not define what he wanted to change from or to. Instead he substituted the mechanisms of reorganizations and elaborate assessment, testing, and evaluation of manager and employee capabilities to cover his inability to define the new TVA regional development mission he claimed to espouse.

The organization Zigrossi proposed divided the work of the group into ten functional and two staff groups , each headed by a vice president reporting directly to President Zigrossi. The ten functional areas were: land management; water management; community partnerships; engineering services; property maintenance; public safety services; senior environmentalist; senior scientist (Muscle Shoals); technology advancements (Chattanooga); and Land Between The Lakes (Golden Pond, Ky). Again all but four of the functional areas related to TVA operations. With this reorganization Zigrossi increased the number of Vice Presidents reporting to him from 3 to 12, which is hardly a way to reduce overhead.

Zigrossi then vacated all senior positions held by current employees in these functional areas and posted the new vice-president jobs in TVA's automated Vacancy Posting System. He employed the Tennessee Assessment Center, a Knoxville company run by two University of Tennessee professors, to screen all candidates for the senior level positions at a cost of $1,000 per screening of a maximum of 300 persons or a possible maximum cost to TVA of $300,000. The tests administered by the Tennessee Assessment Center were described by the Center as 'psychological instruments administered and behaviors observed during the assessment period . . .'. However, the document describing the tests does not state the behavioral standards that were considered normal and acceptable. These tests appear to be more related to attitude than to skills which were needed in the regional development program.

Zigrossi defended this reorganization as refocusing the Resource Group efforts by minimizing overhead and maximizing dollars going to local communities. He did not describe what was to be done to maximize the dollars to communities. Once the reorganization was complete Zigrossi stated that the Resource Group would be better prepared to fulfill its mission and help TVA meet the challenges of the 21st century. He stated 'to be competitive every business must meet the challenge. The key is to figure out a way to balance the economy with the environment. If we meet the challenge in the Tennessee Valley, then we can help meet the challenge on a national and global scale. We must focus our resources on that. Quality will

**Figure 4.9    Norman A. Zigrossi, FBI Agent; TVA Inspector General, 1986–1992; President, Resources Group, 1992–1994; Chief Administrative Officer, 1994**

*Source*: TVA Staff Photo.

be integrated into our overall operation and will be part of every manager's job'. Such general statements did not help to clarify the mission issue. How such generalities could become the basis for reorganization and employee evaluations was not explained by Zigrossi, the hired consultants, or the TVA Board. Some of the TVA employees involved in this testing process have stated to the authors that proven and demonstrated individual abilities as documented in their employment records were never considered in any of the assessments or evaluations for the top management positions. In fact, many felt that experience had become a liability.

Considering the TVA Board's commitment to 'employees first' and the 1992 Strategic Plan mission statement on the workforce environment it is difficult to understand the objectives of TVA management in carrying out the Resource Group reorganization in a manner that resulted in TVA employees with 10 or more years of service believing they were being targeted for 'retraining'.[40] Many of these employees were professionals with excellent training and wide experience in regional and community development. They were placed in the transition program for retraining for other jobs inside and outside TVA. Given that none of TVA's top managers had experience in regional development it should not be surprising that they could not define except in the general terms used by Zigrossi what the TVA regional mission and supporting programs should be. One immediate result was that TVA lost most of the highly professional staff that had given TVA its world-wide reputation in regional development. How such actions contributed to TVA's claimed efforts at improved employee efficiency and productivity was never explained.

The results of the selection process, given the attitude of TVA management, was predictable. Of the 12 newly appointed vice-presidents only three were named from previous positions within the Resource Group and only one remained in his prior position. It is also interesting that not one of the appointees had any training or experience in either community or regional development.

The reorganization also did not solve the problems of funding the regional development program. There were claims that most of the TVA federally funded programs duplicated work being done by other agencies. Particular targets were the work of the National Fertilizer and Environmental Research Center (NF&ERC) at Muscle Shoals and the economic development program. TVA did succeed in having proposed cuts in its federally-funded programs restored. One TVA response was to change the name of the NF&ERC to the TVA Environmental Research Center.

## Crowell and Hayes added to TVA Board

Chairman Waters term on the TVA Board was to expire May 18, 1993 and Senator Sasser started looking for replacements to both the Waters seat and the seat vacated by Runyon. On May 7, 1993 President Clinton announced that he would appoint Craven Crowell as Chairman and Johnny Hayes as director on the TVA Board. Crowell had worked for TVA from 1980 to 1989 before becoming Sasser's Chief of Staff. Hayes was Tennessee's Commissioner of Economic and Community Development and a long time supporter of Al Gore. Both were confirmed by the U.S. Senate on July 1, 1993 and became members of the TVA Board on July 14, 1993.

Both Crowell and Hayes announced that they would spend their first two months on the board to become acquainted with ongoing projects. Crowell stated that during these two months he would concentrate on four areas which he considered important to TVA operations: competitive rates and consumer satisfaction; the safety and management of TVA's nuclear program; environmental leadership and conservation; and workforce excellence.

In mid-September Chairman Crowell announced the new strategic goals for TVA which grew out of a planning meeting of the TVA Board and a group of senior managers. The goals were: to put employees first, recognizing that TVA's core strength is its workforce; to establish TVA's environmental leadership; and to reaffirm TVA's commitment to maintaining competitive rates by not raising additional revenue through rate increases for the next four years.[41]

The first goal, employees first, related to a problem that had plagued TVA for more than 15 years. Shortly after Crowell and Hayes took office Senator Sasser asked the TVA Board to listen carefully to TVA employee concerns about job stability before considering any future layoffs. A TVA spokesman noted that TVA was trying to achieve staff reductions through voluntary resignations.[42] In view of the manner in which the 1992 reorganization of the Resource Group was carried out, in which highly trained professional staff were moved into the transition program for exiting TVA, such a response is difficult to understand.

Moreover, TVA continued to emphasize good management as a key to handling its personnel and program problems. David Osborne, co-author of the book *Reinventing Government* was brought in to speak to key TVA managers in November 1993. His message was that competition is the driving force for good non-bureaucratic organizations; that entrepreneurial organizations first establish a clear sense of mission, get rid of the rules, and then let the managers manage.[43] As has been noted many times TVA leadership in the past fifteen years had not been able to define a clear and specific mission for either the power or the regional development programs. Being competitive and consumer-oriented are operating goals once the mission has been clearly and specifically defined. Instead the employees were subjected to what must have seemed to them to be endless changes in managers and reorganization; a fact apparently not recognized by the TVA Board and its top managers, or the Congress. In response to an employee's question expressing concern that the same managers were to remain in an upcoming reorganization Chairman Crowell stated he had only been in his position two months and that he was not going to dismiss people working directly for him without giving then a chance to prove themselves.[44] Employees of the Resource Group probably wished that the President of that group had taken this approach in his 1992 reorganization.

The second goal, establishing TVA's environmental leadership, is also interesting given the actions subsequently taken by TVA. In less than a month after approving this goal the TVA Board approved a lease of mineral rights it owned in a wildlife management area. One of these leases was to the Sugar Ridge Coal Company. In spite of opposition, TVA supported the company which had a questionable environmental record and even promised that the mined areas would be reclaimed in a model fashion. In the end TVA not only did not receive the royalties called for in the contract but eventually would have to pay part of the cost of reclaiming the land.[45]

The third goal, a commitment to maintaining competitive rates is certainly a sound operating objective given the limited TVA view of the power program mission as being competitive with other private power systems. In the February 8, 1994 issue of *Inside TVA* noted 'The race is on as TVA and other utilities face deregulation. And only the strongest will survive'.[46]

Overall, these goals do not appear to be much different than the objectives established by Runyon. As in the past the new TVA Board did not face the issue of clarifying the specific TVA mission. It outlined the goals listed above but to what end? Would TVA demonstrate for the nation a system for safe, easily constructed and maintained nuclear energy? Would it face up to the key problems of the region and the nation such as waste disposal, clean air and water, and education by trying to work with federal, state, and local governments to find new institutions and programs capable of dealing with these problems? To illustrate, probably less than half of the counties in the Tennessee Valley have either the financial or technical resources to solve their waste disposal problems. While research on these problems might help, it is not an end in itself but must be supported by capable institution and program building.

As it had repeatedly done in the past, TVA turned again in 1994 to reorganization, the seventh in 15 years, as a solution to its problems. On January 6, Chairman Crowell announced a new organization to take effect February 7th (Figure 4.10). Under the new plan TVA would once again return to having a Chief Operating Officer. The Fossil, Hydro, Resource, and Consumer groups would all report directly to the COO. Nuclear would be a separate group reporting directly to the TVA Board. A new position of 'Chief Administrative Officer' was created and the heads of Finance and Administration, Human Resources, Labor Relations, Diversity Development, Public Safety and Communications as well as the Senior Vice President for the new office of Strategic Planning would report to the CAO.

The President of the Resource Group was named as Chief Administrative Officer leaving the future program direction of the Resource Group as a key element in TVA's regional development program very much in question. The major program emphasis for the Resource Group appeared to be to support individual community and area projects. The Morgan Square Project in Greeneville, Tennessee is an example of this project approach to regional development. This new direction of the regional development program stood in sharp contrast to past efforts to encourage new institutions and institutional direction to solve community and area development problems. Perhaps Morgan Square was the kind of project that Zigrossi had had in mind when he stated that the reorganization would 'maximize money going to the communities'.

In its preliminary draft of the environmental assessment TVA seemed to justify its support for the Morgan Square Project as an expression of its commitment to economic development and resource conservation. This was at best a generalized committment which TVA shared with dozens of other federal, state, and local agencies. Moreover this commitment was not unique to TVA as a regional development agency. One would have expected TVA to have had a specific program description to which projects such as Morgan Square could have been related.

As an additional justification for support of the project TVA noted that Sections 22 and 23 of the Act charges the agency 'with fostering the proper physical, economic, and social well-being of the Valley's people'. This interpretation of Sections 22 and 23 differs significantly from the original 1933 analysis by Tracy Augur of the origins of these sections and the powers actually convened to TVA (see Appendix A). As Augur noted these sections were designed to give TVA authority to undertake studies of regional problems in cooperation with the states and their subdivisions but it does not grant any developmental powers to TVA. Nevertheless in 1994 TVA agreed to assist the project by providing a grant of $1 million to help fund the project. TVA then justified the project as follows:

> First, it is needed to rehabilitate and prevent further deterioration of some of the most important buildings in the town's large historic district. Second, it is needed to bring in new occupants to the downtown, which has many vacant buildings,to create jobs and increase the tax base. Third, it is needed as an additional point and lodging to attract tourists to the town who otherwise would pass by and to get those who visit to stay longer. Fourth, it is needed as a first-class meeting facility to attract visitors for conferences and to keep in the local economy the expenditures of local groups which now go elsewhere to hold conferences and activities such as wedding receptions. Finally, it is needed as a source of income and grants for a new local foundation, The Fund for Greene County, which will own the project.

To accomplish these purposes the project 'would rehabilitate four interconnected former hotels, now referred to as the Hotel Bramble, and convert them into the Morgan Square Inn. The inn would have 53 rooms, a 90–seat restaurant, and about 4,000 sq. ft. if retail space. A new conference facility about 7,000 square feet in size capable of seating 400 people would

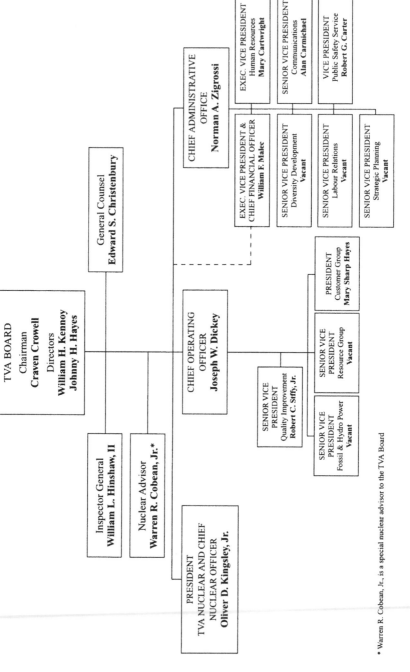

* Warren R. Cobean, Jr., is a special nuclear advisor to the TVA Board

**Figure 4.10  TVA Organizational Chart as of February 7, 1994**

be built adjacent to the inn. The facility could also host receptions and gatherings for up to a 1,000 people. Several outlying buildings would also be rehabilitated for development of about 27,000 sq. ft. of retail and office space. New parking would also be built'.

In spite of the words of praise for the project and its projected impact on the redevelopment of the Greeneville downtown area, the project must be considered highly speculative. The history of efforts to rehabilitate downtown areas, particularly in small towns which generally do not have a strong downtown employment, does not suggest a basis for a good investment. The project was estimated to provide 100 new permanent jobs; not an impressive figure for a county which has had high unemployment for many years.

Moreover the project statement did not indicate that it was a part of an ongoing redevelopment program for the entire downtown area. In fact the project did not indicate strong relations to the National Park Service site, Andrew Johnson tailor shop, and museum and the spring area which was important to the early development of downtown Greeneville and which contains a listing of the early settlers in Greene County.

One unusual paragraph in the Final Environmental Assessment stated that '. . . Congressman Jimmy Quillen enthusiastically endorses Morgan Square, which he calls 'my project'.[47] How such a political endorsement related to the environment of the project is not clear though it may explain the priority given to it by TVA. In a similar approach to economic development TVA announced on Oct. 1994 that it would provide $1 million to help Ijams Nature Center in Knoxville raise $4 million to expand the center.[48] If the Morgan Square and Ijams Center projects did in fact represent the program for the use of appropriated funds, TVA was moving in the direction of providing grants to local projects. There must be hundreds of potential projects which individuals consider important projects for area development. How would TVA choose among them? What the TVA role should have been in offering these grants would have required a more definitive statement of objectives rather than merely a generalized commitment to economic development. But in the uncertain climate of the mid-1990s such a statement never appeared.

## The End of TVA's Regional Development Program

In February of 1997, Craven Crowell, TVA's then current Chairman, proposed to eliminate most of the remaining non-power activities spelled out in the original congressional mandate of 1933. Crowell's motivation in making his proposal was no doubt to dampen self-serving assertions by TVA's private power competitors that TVA enjoyed unfair subsidies in an era of growing deregulation and competition. Crowell's proposal was, in effect, a sop to those conservative Congressmen who never had much use for TVA in the first place. These cuts would have eliminated flood control and navigation responsibilities as well as TVA's few remaining economic and environmental development programs. It would have left TVA as simply and one is tempted to say, merely, a wholesale power producer, owned in significant measure by private shareholders through its bonded indebtedness, superimposed on an enduring but lessening investment of the 'People of the United States,' made through tax dollar appropriations. Crowell's proposed cuts would have resulted in the elimination of another 2,000 jobs, (though some of TVA's non-power activities would under the plan be undertaken by contract with or shifted to outside suppliers and service providers). This downsizing step would in effect have completed the dismantling of TVA as a integrated regional development and planning agency. Ironically, this final elimination of any claim to regional development would have had the effect of paving the way for more, not fewer, attacks by private power interests on the agency's residual public status. With the last fig leaf removed, there would be no rationale whatever for TVA to make a claim for its status as a public authority. Crowell's proposed radical changes were not implemented as he had laid out. But in 1998 changes in TVA's Congressional budgeting resulted in changes for the regional development program that were nearly as drastic.

The final death knell of the TVA regional development program was sounded in mid-1998 when TVA Chairman Craven Crowell told Congress that TVA no longer needed to receive its annual appropriated budget from Congress. After years of declining ability to define what its regional development program was, TVA found that the program had become more of a distraction than an asset. Crowell had stated publicly that the relatively small percentage of the total TVA budget represented by the appropriated funds, and the activities they funded, required a disproportionately large amount of his time. Also, at a time when deregulation of the electric power utility industry was under heated discussion and was expected to become a reality, TVA faced increased criticism from its competition, the investor-owned electric utilities, that the appropriated budget was subsidizing TVA's power program. Even though this was not true and was explained repeatedly, the private power companies kept up the assertion as a competitive move. Crowell indicated that he decided to forego the appropriated funds so as to remove the subsidization issue from the deregulation discussion. The result of this retreat was the termination of the regional development program after 65 years of regional planning activity within the Tennessee Valley Authority. TVA had become, more or less, just another power company.[49]

## Notes

1    TVA *Annual Report*, 1980
2    *Newsweek* September 3, 1979, p. 35
3    *TVA Today* April 1979, p. 2
4    A. J. Gray interview with John Bonum, August 12, 1988

5   Memorandum, Sharlene P. Hirsch , Manager, Office of Community Development to S. David Freeman and Richard Freeman, April 23, 1979 entitled 'Status Report on Office of Community Development: The First Two Weeks'
6   *Knoxville News-Sentinel* June 27, 1979
7   *Knoxville News-Sentinel* September 26, 1979
8   Tennessee Valley Authority, 1980, *Strategies for the 1980's; A TVA Statement of Corporate Purpose and Direction, 1980*
9   *ibid.*
10  *Inside TVA* December 15, 1981
11  *Knoxville News-Sentinel* January 27, 1981
12  *Knoxville News-Sentinel*, January 20, 1982
13  Tennessee Valley Authority, *TVA: An Agency for Regional Development*, 1982
14  *Knoxville News-Sentinel*, July 11, 1983
15  Tennessee Valley Authority Merec Staff, *Managing Energy and Resource Efficient Cities from Concept to Impacts, A Report to the United States Agency for International Development*, Knoxville, TN, April 1988, 48 pp.
    Also, Tennessee Valley Authority Merec Staff, *Managing Energy and Resource Efficient Cities (MEREC) Summary and Evaluation, 1981–1989*, A Report to the United States Agency for International Development, Knoxville,TN, September, 1989, 31 pp.
16  *Knoxville News-Sentinel*, April 22, 1984
17  *Inside TVA*, May 1, 1984
18  *Knoxville News-Sentinel*, April 30, 1986
19  *Inside TVA*, December 10, 1985
20  *Knoxville News-Sentinel*, April 28, 1985
21  *Knoxville News-Sentinel*, May 11, 1986
22  John Kenneth Galbraith, *A Journey Through Economic Time*, Houghton Mifflin Co. Boston and New York, 1994, p. 176
23  *Knoxville Journal*, August 4, 1987
24  *Inside TVA*, January 20, 1987
25  *Knoxville Journal*, July 5, 1987
26  *Inside TVA*, February 2, 1988
27  *Knoxville Journal*, February 16, 1988
28  *Inside TVA*, April 27, 1988
29  *Knoxville Journal*, April 15, 1988
30  *Knoxville Journal*, May 19, 1988
31  *Inside TVA*, April 24, 1990
32  *'Labor-Management Relations: Tennessee Valley Authority Situation Needs to Improve'* U. S. Accounting Office, June 1991
33  *Knoxville News-Sentinel*, August 20, 1991
34  *Knoxville News-Sentinel*, December 15, 1990, November 2, 1991, and July 16, 1994
35  *Inside TVA*, January 15, 1991 and January 14, 1992
36  *Knoxville News-Sentinel*, June 11, 1992 and January 27, 1994
37  *Inside TVA*, April 7, 1992
38  *Inside TVA*, August 11, 1992
39  *Inside TVA*, December 1, 1992
40  *Inside TVA*, November 17, 1992. In addition, Zigrossi in a May 30, 1993 interview with the *Knoxville News-Sentinel* made the observation that some Resource Group employees 'who were valuable people 10 years ago under the new thrust of the organization aren't valuable assets . . .' On June 8, 1993 in a letter to all Resource Group employees he retracted the statement saying his comments 'were directed to programs over 10 years old'. Since Zigrossi had never described in specific terms, what the thrust of the new organization was what he really intended to say is not clear.
41  *Inside TVA*, September 24, 1993
42  *Knoxville News-Sentinel*, July 27, 1993

43  *Inside TVA*, November 30, 1993
44  *Inside TVA*, October 1993
45  *Knoxville News-Sentinel*, October 11, 1993 and March 17, 1994
46  *Inside TVA* described the problem in its December 15, 1981 issue
47  *Final Environmental Assessment, Morgan Square Project, Greene County*, June 1994, Prepared by the Tennessee Valley Authority, p. 3
48  *Knoxville News-Sentinel*, Oct. 27, 1994
49  *Knoxville News-Sentinel*, July 27, 1998

# Chapter 5

# Some Reflections on the TVA Experience as a Regional Development Agency

Martha Derthick in her book *Between State and Nation* discusses regional development organizations in the United States and makes the observation that:

> They may help resolve interstate conflicts or promote interstate cooperation. They may enable the federal government, a group of states, or the federal government and the states together to pursue public purposes for homogeneous areas that do not conform to established jurisdictions., They may also be a means of decentralizing federal administrative activities and improving coordination of such activities. How well regional organizations serve these objectives, how they may be improved, and whether to extend then are at issue.

> The distinctive virtue of regional organizations is that they are suited to respond to particular needs or problems assailable on a regional scale and somehow peculiar to an area as a natural or social or economic unit.[1]

Our investigation of TVA's experience over some sixty years seems to support the views outlined by Derthick. We strongly support her view that regional organizations are suited to respond to specific area needs or problems. One problem with such agencies is that in many cases the legislators or the agency itself have not been able to describe these needs and problems in specific operational terms. As a result the agency turns to generalities such as job creation or economic development to describe its mission. This was certainly true of TVA during at least the last thirty years.

This investigation also suggests that there has been not one but several TVA's. For example the TVA of the period 1953 to 1973 was quite different from the TVA of the 1930s and 40s. The TVA of the period 1974–1994 did not even remotely resemble the TVA of the two earlier periods. This fact about TVA has not been recognized by either the friends or foes of TVA. To refer to TVA without recognizing the particular stage in its history does not permit more than a general comment on the agency.

Thus when S. David Freeman talked of taking TVA back to its roots his statement lacked meaning in the absence a recognition of where TVA stood at that time which could indicate the direction needed for a definition of a new and urgent mission. When Craven Crowell called for renewing the pioneering spirit within TVA he needed to relate such a call to a recognition of TVA's then current limited programs as a basis for the direction of change needed as a part of a clear and specific new mission for TVA.[2] But perhaps what is more significant as an indication of the evolving direction of TVA is the type of appointments made to key staff positions. For example the appointment of Alan R. Caron, a utility marketing executive with 23 years experience in developing marketing strategies for the electric and gas industry, to the post of Senior Vice-President for Strategic Planning with responsibility for positioning TVA competitively in the utility industry was an indication of the probable limited scope of the future TVA regional development program thereafter.[3]

But there are other issues regarding regional agencies which should be addressed. Central issues that affect regional agencies fall into four closely interrelated categories all of which bear on the ability of the regional agency to operate effectively: (1) organizational pressures resulting from the completion of the original mission; (2) how the agency deals with changes in demographic and economic conditions which change the nature of the problems facing the region; (3) how the changing mission of the regional agency is recognized within the agency particularly when the new mission is likely to be radically different from the original mission and has to relate to changed regional conditions; (4) how a regional agency becoming essentially operational should transfer non-essential operations to other agencies. TVA has had to face each of these issues over the past thirty years. It still faces them.[4]

### Implications of Completion of the Original TVA Mission

The strength of the original TVA mission was that it outlined specific duties that were sharply focused on the Tennessee River Basin. It then superimposed on these specific duties a general assignment (Sections 22 and 23) to be concerned with the physical, economic, and social resources of the area in cooperation with state and local governments.

The resource problems in the region were well known and generally understood at both national and regional levels: soil erosion; forest deterioration resulting from annual burning and poor management; over half the population living on farms with subsistence rather than commercial farming typical of the region; poor health and other community services; deteriorating fish and wildlife resources; and cities unable to provide essential services because of low revenues and limited technical services.

The three original board members although men of different training and temperament and disagreeing on some key issues really complemented each other: A. E. Morgan was an engineer with experience in multipurpose river development and in area development; H. A. Morgan was an agronomist with experience in the workings of the state agricultural colleges and the county agent system; and David E. Lilienthal was a lawyer and a former state commissioner of public utilities. Each had an established professional reputation. It is true that there were board disagreements among the three over how to handle the TVA power program but in general there was agreement on the overall thrust of the general regional development program. They agreed on the problems facing the region, that an expert staff was needed to deal with a broad range of area problems, and the need to relate the proposed reservoir system to the overall regional development program. Thus the staff included experts in agricultural, forestry, fish and wildlife, water control in the channel and on the land, public health, recreation, and community and regional planning.

The original board also seemed to understand, by instinct or by consideration of the options, that not all problems could be solved by limiting TVA activity to the Tennessee River watershed. The watershed made little sense as a region for power generation and distribution. They all accepted the seven valley states as a more logical region for resource development because the original board supported the idea that new institutions and programs were needed to deal with these problems. TVA seems to have been the only regional agency to recognize the importance of using different regions for different type of regional problems. It is true that the later-established Appalachian Regional Commission (ARC) required states to create multi-county local development districts but these were not the kinds of broad interstate regions needed to handle many of the problems which ARC tried to address. Moreover, ARC was handicapped by relating to parts of states rather than to entire states within its area of concern.

Both the TVA Board and staff seemed to expect the region to change as river improvements were made and as progress was made in dealing with the multiplicity of regional resource problems. One result was that TVA adopted the policy of shifting to the state or local governments those TVA programs that had become operational at those levels. For example the TVA forestry program changed its emphasis from fire control to reforestation, to forest management, to tree genetics as the nature of the regional forestry problems changed and the operational responsibilities formerly shared by TVA were taken over by other levels of government.

All of TVA's regional development activities were related in some fashion to the river improvement program. TVA's budget for regional development was justified in part by its support of the river improvement program. As the system of dams included in the 1936 plan for the improvement of the Tennessee River was completed the problem faced by TVA was how to justify a continuation of the regional development program. This was a difficult time for TVA because it required a complete reassessment of area problems and possibly an opening of the original TVA Act to amendment; something that TVA had strongly opposed in the past because it could not control the outcome. Instead, TVA tried to reorient the regional development program within the framework of the original Act and without the benefit and support of new specific duties related to new regional problems.[5]

The need to adapt to changes in the region raises some difficult question for regional agencies which have completed their original mission. Should the original legislation for such agencies contain a provision for termination or reassessment once the specific duties assigned to an agency are completed or no longer relevant to the region. The TVA act, of course, did not contain such provisions.

## Changes in Regional Problems

The TVA experience is an example of how, as a region changes over time, so does the nature of its problems. As noted earlier in this study, during TVA's first 30 years the Tennessee Valley and in fact the entire South changed dramatically from an agrarian subsistence economy to an urban industrial economy. For example, during this brief period the urban population of the seven valley states grew from 25.2 percent of the population in 1930 to 43.2 percent in 1960. During the same period the non-farm population outside cities grew from 19.9 percent to 36.6 percent. In contrast the farm population declined from 54.9 percent to 20.2 percent. The region was well on its way to a population distribution and an economy that began to approach the national averages.

Regional needs and regional problems were also changing. In general, natural resource problems were being brought under control and the region's forests and agriculture were becoming productive contributors to the economy. While these needed some attention, new problems emerged related to urban and industrial and commercial growth. Air and water pollution, water supply, waste disposal, and problems relating to urban sprawl were becoming increasingly important.

Urban and non-farm employment trends signaled that the problems of the poor and isolated rural areas could no longer be solved except in the context of the new urban trends. But TVA simply continued to give major program emphasis to the same resource problems for which it had been created to deal with. Its new vehicle, the Tributary Development Program, tried to work within the context of small watersheds on the false assumption that what had worked for the Tennessee River Watershed as a whole could be duplicated in its smaller tributary watersheds. It assumed, for example, that by correcting the remaining erosion problems in the Beech River watershed it could solve the economic problems of the area. By building dams in the tributary watershed it assumed it could repeat the contributions the system of main river dams and the related regional development program had made to the region as a whole. As Melvin R. Levin had pointed out in his 1968 article, *The Big Regions*, a watershed is not a satisfactory region for economic development; a conclusion TVA itself had reached in adopting the seven valley states as the region for institution building to solve both the resource and economic problems of the Tennessee River watershed (Levin, 1968).

One can only conclude that neither TVA nor the Congress has been able to make needed changes to meet the new needs and problems now facing the Tennessee Valley. In fact we know of no examples of multi-state regional agencies which have been able to make such an adjustment. On a smaller scale the one example which comes to mind is the Twin Cities Regional Planning Commission which after identifying the specific regional problems in the Twin Cities area worked to create a new agency to plan and operate programs to solve the problems. As late as 1994 the Minnesota legislature further consolidated the responsibilities of the council so that it could better perform the duties assigned to it.[6]

If this kind of action is not taken as a result of such changes, does the regional program risk becoming an anachronism and a bureaucracy trying to perpetuate itself? Certainly TVA's efforts to define the regional development program and its repeated reorganizations did not solve the problem of meeting the needs of a changing region. As we have noted earlier in this study it might have been a wiser step for Congress to recognize that TVA had become essentially a publicly-owned electric utility and could operate an economic development program similar to that of private utilities. It could then have turned over to other state and local agencies such activities as managing the recreation areas under its control. As operations they added little to a regional development program and in fact could fit well into existing state and local programs. Much the same could probably said about the Appalachian Regional Commission (ARC). We now have a much more integrated national economy than existed when these agencies were created. Certainly the poverty, unemployment, health, and education problems are now spread across the entire nation and must be recognized and treated as national urban problems.

**Effect of Changing Mission on the Regional Organization and Regional Configuration**

One reason it is important to identify in specific terms the new mission of an agency is to identify the region appropriate to the problems the agency is trying to solve. The idea that one region can be associated with all area problems has not in our opinion proven to be workable. In turn, the type of agency and its organization should be related to the problems assigned to it. These issues certainly should be addressed by both TVA and the Congress. The many reorganizations of TVA undertaken without first defining the agency mission did not solve the problems of employee morale or of the nuclear program. In fact the nuclear problems probably reduced the much needed attention required by the regional development program. In turn, trying to hold onto an undefined regional development program undoubtedly prevented consideration of the kind of organization and program needed to keep the nuclear power program on track.

From the generally available information on recent TVA programs it is not clear what region TVA is currently trying to serve and work with. In the discussion of the 1992 reorganization of the Resource Group, TVA management referred to TVA's helping to meet both national and global challenges. If this is true, then perhaps a clearly defined demonstration program serving the needs of the entire nation is called for.

## A Regional Development Agency as an Operational Agency

As TVA completed its original mission it did make an effort to redefine a new mission for its regional development program. The effort to use the tributary area development concept (TAD) to create new regions for development is an example. This effort was not really successful in meeting the demands of the new broader regional problems and gradually declined in importance until its merger into the community development program in 1979. In the years that followed neither the board nor top management was able to clarify and give the leadership needed to articulate a new direction to the regional development effort; even though there were many general statements under the a variety of labels such as 'the strategic plan for the future'. As a result all non-power activities became grouped under programs financed by Congressional appropriations. This combined all kinds of operations such as reservoir and shoreline management, and overall agency security. Over the years these operations consumed more and more of the non-power budget as TVA continued to support operations that in other times TVA would have turned over to states and localitiess. Examples are the reservoir recreation areas still managed and operated by TVA. A precedent for such transfers occurred with the conveyance of the extensive Land Between the Lakes reserve to the US Forest Service.

In the 1995 budget only $55.5 million was scheduled to be used in the regional development program, with the remaining budget allocated to water quality monitoring and operation of the Tennessee River system, for water and land management, and, of course, the power program. TVA today is primarily concerned with operations, especially power operations, raising the critical question, is the TVA Act with its broad mandate for regional development still the appropriate legal authority for such operations?

## General Observations

The 1940s and 50s was a period of widespread interest in the special problems of what then appeared to be distinct regions in the United States. Certainly President Franklin Roosevelt's description of the South as the nations major economic problem is clearly no longer valid. In the past 60 years the nation has changed dramatically to the point that there is now a truly national economy; the economic differences between the nation's geographic regions have been reduced. This raises the question of the need for large multistate regions as contrasted to nation-wide programs.

In this kind of situation is it possible for TVA to articulate a new mission that is unique to a regional agency? It is questionable if today TVA's top management could devise such an innovative program as the flood damage prevention program of the 1950s and 60s. In that period TVA had board members with broad interests and experience. TVA also had a multi-faceted professional staff which had access to the board so that it had a broad range of views to consider in developing policy and approving programs. The present driving forces of competitiveness and consumer satisfaction, while good operational guides for the power program, do not provide much substance for guiding a new and revitalized regional development program, assuming such a program is needed.

This problem involves not only TVA. It concerns the basic idea of regionalism in the United States. The regional concepts and institutions of the 1930s, 40s, and 50s have served us well; but do they meet present day needs for a national urban policy and programs that can deal with the problems of the inner city, the city-suburb conflicts, and the system of urban settlements. The demographic and economic changes of the past 30 years require a new look at the problems of creating new institutions for the delivery of urban services. For example, in the Tennessee Valley and in other parts of the country, many small, poor, and thinly populated counties cannot provide adequately for such basic services as education and waste disposal. Other institutions and programs at the state and local levels also are needed to deal with these problems and other such as water supply, and water pollution. These matters should be a part of the re-evaluation of regionalism in the United States.

Today, TVA faces an identity dilemma. It is possible that TVA will not survive in its present form into the 21st century. In fact, in some important ways, it has already *not* survived. When TVA was forced, during the Eisenhower era, to accept self-financing of the power program TVA ownership in effect moved from 'the People of the United States of America' to the bondholders. This action moved TVA in the direction of being a private corporation so that by the 1990s TVA was barely distinguishable from a private power company. Instead of stressing its public functions TVA stressed its corporate functions and increasingly its self-identification became dominated by corporate images rather as an agency with a public purpose. New logos, endless reorganizations, new corporate titles for top management, expensive and repeated corporate office decorations, public relations campaigns instead of mission re-identification and renewal of purpose, low staff motivation and morale, the selection of staff on the basis of attitude rather than skills, misreading the social and economic data to conclude that 55 percent of the Valley's population live in rural areas and on this basis decide to create a center for rural studies. Even the Technical Library which had kept the

TVA technical staffs informed of the latest technical information in their professions became 'the Corporate Library'. TVA's current TV advertising campaign directs power users to call on TVA or their 'local power companies' for advice when in fact there are no local power companies. Locally, TVA power is distributed by municipalities and cooperatives (though this is likely to change in the near future).

All these actions bespeak an agency in confusion and distress, a pale shadow of the TVA of the 1930s, 40s, and 50s. In fact today TVA does relatively less in economic development planning and conservation assistance than its private counterparts to the east and the south. TVA, moreover in the mid-1990s is regarded as a major malefactor in the eyes of the environmental community. This is a sad commentary for an agency that was once the nation's leader in regional development and conservation planning. In the turmoil within present day TVA it is clear that TVA has forgotten its institutional history and in so doing has lost its innovative edge.

Does TVA have a future? TVA will no doubt survive in some form but its future could take one of several directions. It could muddle on, be abolished by Congress, or it could be given a new and vital mission that could make it valuable and needed by the nation. It could concentrate on the single mission of a power producer but this raises the issue of the public purpose of the publicly-owned power system. As we have noted earlier, merely being competitive is not in itself a sufficient public purpose. Competitiveness is an operational objective which it shares with private power companies.

TVA could drift on into the future more or less as it is doing today. Federal appropriations, apart from power income, are now almost invisible when compared to the total TVA budget, and could disappear altogether as a new Congress sees TVA grants as pork barrel for a particular region and the Valley no longer the basket case of the nation. Even though cities and related urban problems dominate the nation and valley concerns, TVA is still treating the valley as a rural area. An example previously noted is TVA's funding a new rural studies center, claiming that 55 percent of the people of the Valley live in a rural setting, a figure which is not consistent with census data. In any event, the elimination during the early 1990s of most of TVA's planning and development staff has left TVA without any real capability in regional development, urban or rural.

Congress, of course, could simply abolish TVA as a public agency. The consequences of this action would be difficult to assess in light of the long history of TVA service to the public good. But the question has to be asked; would a totally private TVA be much different from what it is today?

There is conceivably another potential future for TVA, one which under a different name would have a new national demonstration mission to which Sections 22 and 23 of the TVA could be related. This would be for a TVA-type agency to become, once again, a development and conservation agency for the southeastern states doing what neither states and localities nor the single purpose federal agencies could do, namely, linking private and public sectors in a creative partnership to deal with overall development and particularly the urban crisis in America. It would attempt to link the environmental, transportation, jobs in the inner cities, communication, and energy issues to the emerging pattern of cities, towns, and villages. This would provide a wide choice of urban living patterns and at the same time support open space between the urban centers for food and fiber production. It would relate such development to a flood damage prevention program so as to prevent such problem as occurred on the Missouri in 1993. Such a program would also relate recreation, scenic, and fish and wildlife resources to a regional development concept and program.

This is a dream built on the early foundations of TVA. A new constituent unit separate and independent of the power part of the agency could become a national demonstration agency to find new ways to address the difficult, critical, and seemingly intractable, but interrelated, development problems urban, suburban, and rural America. The National Resources Committee noted that 'the larger import of TVA . . . is national in its scope'. It quoted one of the original directors of TVA as stating that 'the national purpose of this test (i.e., TVA-ed.) is continually in the mind of the Board of Directors. Every step taken, each project set up, each result obtained is weighed from the point of view of its possible application to other parts of the country'[7]

What issues or problems could this new TVA-type agency address? Nothing less than the burning issues facing the nation as a whole:

Untangle the knotty maze connecting energy, environment, communications, and the economy to move closer to levels of sustainability;

Deal with the problems of decaying cities and wasteful, sprawled development in the countryside;

Provide training and jobs for the unskilled;

Make the information highway an integral part of both regional and national development for the benefit of all sectors of society, not merely the already information-rich;

Develop and introduce new modes of electric-based transportation into urban areas;

Adapt and integrate into the development process alternative, more benign, ways to produce energy such as solar and wind power.

The list could go on. The nation has many interwoven development problems. But whatever its agenda, the new agency would have to have a commitment to its role as a national demonstration of how to deal with the country's growing economic and environmental problems. It would also need a federal mandate to undertake such a program.

Is it likely that TVA could once again become a dynamic demonstration agency? Probably not – agencies and organizations are like living organisms; they are born, they are energetic and lively in youth, stable in middle age, atrophy and become forgetful in their maturity; they struggle on thereafter, often losing sight of their original purpose and unable to respond to new and different problems. They rarely succumb, but rarely do they reinvigorate or reincarnate themselves. What makes the reinvigoration even more unlikely are the limitations of and constraints on the TVA Board and top management. There is no longer an A. E. Morgan, a David Lilienthal, or a Gordon Clapp to lead the way. So it is probable that TVA will muddle along in the 21st century, remaining primarily a power producer so concerned with operations that it is not capable of considering the broader public purposes of its power activities. It will probably give lip service to a regional development, a program long since dead, and leave the Southeast and the country wondering what it was all about in the first place. Perhaps this review of TVA's regional planning and development record will help to remind us of what TVA once was – and what it could have remained had it acquired and adopted the capacity to meet the problems of a changing world.

We believe the problems facing TVA as a regional development agency apply as well to regionalism as practiced in America. Regional concepts and institutions need restudy and appraisal. The demographic and economic changes of the past 40 years require that we look again, and in a new way, at American regional institutions. We believe new institutions and programs are needed at state and local levels to deal with such problems as waste disposal, water supply, air and water pollution, and education. These are a part of an overdue re-evaluation of the past – and the potential future – of regionalism in the United States.

## Notes

1    Martha Derthick, *Between State and Nation: Regional Organizations of the United States*. The Brookings Institute, Washington, D. C. 1974
2    *Inside TVA*, November 1, 1994
3    *Inside TVA*, October 18, 1994
4    For the most part comments on these issues refer to TVA and similar types of regional agencies such as the Appalachian Regional Commission (ARC) and the Regional Economic Development Commissions which were designed to cover other parts of the United States outside Appalachia. Other interstate regional groupings such as the Southern Growth Policy Board provide forums for advice and discussion to state governments and are not comparable to TVA and similar operational development agencies.
5    TVA is not alone in responding in this manner to the completion of the specific duties assigned to an agency. The Appalachian Regional Commission (ARC) is still facing this problem and as a result, almost yearly faces the threat of congressional rejection of its regional development program
6    'Super-agency Replaces Twin Cities Metro Council' *Planning*, July, 1994, p 30
7    National Resources Committee, *Regional Factors in National Planning*, 1935, p.83

# Postscript

The role of planning in a democratic society is often portrayed as in conflict or tension with market-based decision making. But as there are no pure or perfect markets, there is no perfect or total planning. Societies can, of course, drift, making decisions on an emotive or ritualistic day-to-day basis. But most successful societies and economies have sought to bring the incremental decision process under some generally accepted rational framework. The planning function has had a place in every society in human history.

Planning has traditionally served two important societal functions:

The first function has been to address externality problems that arise due to market imperfections or to information limitations. Externalities can be direct costs which can be shifted to the larger community by the market, or they can comprise opportunity costs – losses attributable to piecemeal or uncoordinated decision- making. Simply drifting from day-to day in the development process can inflict great costs on communities and societies. Failing to integrate connected decision arenas can result in significant lost opportunities.

The second function of planning has been to give voice to future generations, whose interests tend to be excessively discounted by the short term orientation of the market.

A third function has been added in recent decades by planning theorists who have come to the view that capital constitutes more than money capital or fungible assets. Capital includes the inherent value, however measured, of environmental capital -- the air, water, land, habitat, species diversity and so on. Capital accounting also now includes social capital: the value of and investment in people. An accounting system that is solely focussed on economic assets or flows is a partial system, encourages the exploitation of natural and human capital and their transformation into economic, monetary capital. Thus, planning has an obligation to identify a suitable balance that must be struck among these forms of capital for systemic sustainability to prevail. This requires developing an understanding of how economic, social and natural capital interact with and relate to each another. Finding this balance is the foundation of what has come to be called Integrated Resource Management. Planning policies and projects thus can be judged according to how well they fulfill these three functions. A key question this study has raised is: how well did TVA do in its efforts to address these functions through integrated resource management? The answer, as we have seen is a mixed one. Arthur Morgan's grand vision to re-make industrial urban society was clearly at odds with David Lilienthal's notion of bringing low-cost electricity to the South to spur development and attract jobs. Both views utilized what we might call 'planning imagination'. Arthur Morgan's view of the Good Society was essentially anti-urban, emphasizing the self-sufficient small town, a romantic dream that FDR may have shared. But the encompasssing trends were in the opposite direction – toward urbanization and increasing scale, as Lilienthal clearly understood.

Electrification of the Valley inevitably became the centerpiece of the TVA experiment until it ultimately became the dominant activity of the agency. By 1960 the hydroelectric potential of the Tennessee River and its tributaries had been fully realized, yet demand for electricity continued to grow. By 2001, hydro power supplied merely 6 percent of TVA's total output, whereas coal furnished 65 percent and nuclear, 29 percent. The river has long ceased to be the organizing principle for the activities of the TVA. Thus the definition of TVA's 'region' has shifted or blurred. So the question is just what is the TVA region today? Is it the actual real estate controlled by TVA, limited to its shoreline properties and the waters and river they contain? Is it the power distribution region to which it wholesales electricity on an exclusive basis? (TVA has had a monopoly to supply power within a defined region and beyond which it may not go. But with the deregulation of energy and the long distance transmission of power, this limitation to a geographic power region may be ending.) Is it the aggregration of the seven southern states, portions of which are in the river basin? Or is it the nation at large, for which the TVA serves, or could serve, as a demonstration of good practices and a source of power for national defense or emergency needs?

TVA has served all of these regions in varying degrees. But its regional focus and emphasis has changed over the decades since its creation. As we have seen, TVA's internal planning function has ebbed and flowed in response to changing perceptions of the role of the agency and the region it serves. Thus throughout its history planning within TVA has been an appendage to the operational functions of the agency rather than an overarching and directing feature within which operations are subsequently pursued. But as this study has shown, the planning function  has not been inconsequential within the agency. Planning accomplished much that was valuable and occasionally provided models for the rest of the nation to follow. Examples would include the national flood insurance program modeled after TVA initiatives and the 701 Local Planning Assistance program which provided planning and development assistance for communities across the country. Strengthening of State Planning departments and fostering of state policies on parklands, industrial development, and recreational facilities were all aided and informed by the work of TVA planning staff.

The planned developments at Norris and Timberlake at Tellico have been called failures by critics. But in fact they were not. Norris never realized the dream of Arthur Morgan or FDR for new communities connected to the electric grid and the larger economy blanketing the Tennessee Valley. But Norris demonstrated the value of foresight and community planning in creating a superior living and working environment. Timberlake was never built in the fashion that the TVA staff had envisioned but after the site was sold, its development as a planned waterside community followed some of the ideas set forth for the planned new town. But in retrospect it is clear that had the planned TVA new town been carried out, the result would have been superior to the development that subsequently took place.

TVA planners had little choice but to be flexible and pragmatic rather than comprehensive and integrative. Their powers and influence within the agency were always subordinate to the specific operational mandates Congress set forth in 1933. It is remarkable that they accomplished as much as they did. Today, planning has ceased to be a significant feature of TVA. The management of land is restricted to modest shoreline controls and to the selling off of what are regarded as surplus lands around dams and power stations. But TVA cannot escape the linkage between power generation and environmental sustainability. So land-use decisions, air and water quality, and habitat preservation – along with the traditional unending quest for economic development – are inextricably connected to TVA's functions, just as they are for private utilities. TVA, however, has a special obligation to show leadership in this search for a balanced, sustainable approach to development. The need for rational, informed regional planning has never been greater – nor more difficult.

## The Regional Resource Stewardship Council

In 1998 TVA lost its non-power funding as part of a deal with Congress. This loss resulted in an orphaning of the regional stewardship function that had long characterized TVA's relation to its service area. Concern for the de-emphasis of the regional support and non-power functions, the rationales, as we have seen, for the very existence of TVA, resulted in efforts by a new TVA Board to reach out to constituencies and stakeholders in the seven-State region. TVA leadership felt, justifiably, that it needed to communicate better with the public. A regional advisory committee was set up by TVA, following the guidelines of the Federal Advisory Committee Act, This committee has since been transformed into a 'Regional Resource Stewardship Council' (RRSC) to provide advice to the TVA Board on setting priorities among competing objectives and values. TVA's stewardship activities include the operation of its dams and reservoirs, its responsibilities for navigation and flood control, and the management of the lands in its custody, water quality, wildlife, and recreation. The Council is purely advisory and consists of up to 20 members appointed by TVA for a term of two years. Seven members are to be nominated by each of the seven Governors of the states served by TVA with four members representing distributors of TVA power, and at least one member serving, respectively, navigation, flood control, recreation, and environmental interests.

The Council has met perioodically in 2002 and 2003 focussing its attention primarily on recreational issues and lake levels. Water quality and public land disposal issues have also been on the agenda. Interestingly, the controversial issue of regional air pollution has been kept off the table, at least for now. Habitat and biodiversity protection, high in environmentalists' priorities, has been voiced by members of the Council, but has yet to be addressed.

It remains to be seen whether the RSSC will be an effective mechanism for restoring a salient regional perspective to TVA decision-making, or whether it will simply amount to a public relations process. So far, the TVA Board seems to have taken seriously RSSC recommendations offered to date, but the difficult issues have yet to be broached. And without a regional studies and planning staff, it is questionable whether the technical capacity to deal with complex development issues is now available within TVA. The creation of the Regional Resources Stewardship Council is certainly a step in the right direction. But time will tell whether its impact is significant. (See Appendix for Charter of the Regional Resource Stewardship Council.)

## Conclusion

The regional planning function within TVA was, from the beginning, subordinate to the three activities specifically defined in the original TVA Act: navigation, flood control, and hydropower generation. Regional planning, that is, the comprehensive physical structuring of the region, its land uses and transportation, was never feasible within the confines of the Act. Moreover, TVA's 'region' was never a single jurisdiction, as each function embraced a different service area.

The community planning function was even less a central function of TVA. Politically sensitive, legally ill-defined in the TVA Act, community planning was necessarily an auxiliary and secondary function that evolved on an ad hoc basis as needs and opportunities arose. The two major efforts of the TVA planning staff to create full-blown new towns – Norris in the 1930s, and Timberlake in the 1960s, were never fully realized. In both cases the mandate for TVA to intervene in urban development processes was weak. Politically it was risky, and their timing vis-a-vis potentially supportive federal programs turned out to be too late to take advantage of funding and loan programs which lost support in Congress. Nevertheless, both new town projects ultimately had beneficial impacts on the local development patterns.

TVA's planning staff had to work with the river functions staff in complementary ways. They had also to work in harmony with State and local planning and development agencies. They needed to be sensitive to local needs as well as political realities. That they accomplished as much as they did is a testimony to their technical skills and acumen. In the end, though, planning assistance to states and localities became a casualty of the evolution of TVA into primarily a power-generating agency. The legacy of the TVA planning effort endures, however, in the form of state parks, barge terminals, industrial parks, and refurbished downtowns along the river. It also continues in the form of enhanced state planning agency capacities, information databases and surveys, and legislative innovations such as flood insurance programs. TVA's urban and regional planners can look back with pride on these accomplishments. The Tennessee Valley is a better place today thanks to their work.

# Appendices

**MEMORANDUM**

February 6, 1943
To: L.L. Durisch and Tracy B. Augur
From: Mr. Howard K. Menhinick, Director, Department of Regional Studies

ORIGIN OF THE REGIONAL PLANNING AND DEVELOPMENT CONCEPT IN TVA LEGISLATION

The TVA Act of 1933 provides for two related programs of activity: (1) The carrying out of a multiple purpose river development, power and fertilizer program, and (2) the inauguration of broad studies aimed at the later initiation of other programs of economic development intended to round out the first.

Both of these programs had been in evolution over a long period and in their later years had had the devoted sponsorship of two great and public-spirited men, George Norris and Franklin Roosevelt. It was only natural that the two ideas should be brought together when Mr. Roosevelt entered the White House.

The legislative history of the river development, power and fertilizer sections of the TVA Act is well known. These programs had been brought together under the sponsorship of Senator Norris in a number of so-called Muscle Shoals bills in the period following World War I. The history of the regional planning and development sections is equally interesting. It had roots in the conservation movement that first gained recognition through the efforts of Gifford Pinchot and Theodore Roosevelt. It had roots also in the programs of land planning for cities and rural areas that paralleled the growth of the conservation movement. The ideas of conservation and land planning for the good of the general public became well developed in New York State during the Governorship of Alfred Smith, and were more fully developed by Franklin Roosevelt in his four-year term from 1929 through 1932.

The first comprehensive plan for the development of a larger region embodying both urban and rural problems was the plan for the State of New York, prepared during the administration of Al Smith. Technical direction for the preparation of this plan came from a group of men who had been associated with both the conservation and city planning movements. Benton MacKaye was one of them, who had worked with the Forest Service under Pinchot and had helped to develop a plan for the resettlement of returning soldiers on the land after World War I. Clarence Stein, an architect and progressive city planner, was director of the New York State Plan. Also associated with the New York State Plan were leaders in improvement of rural conditions. This plan had a great influence on the thinking of Governor Franklin Roosevelt.

In August 1929, Governor Roosevelt proposed a broad survey of land utilization in the state. In the course of an address on this subject he said, "I have long been interested in the general subject of city and of regional planning. The present proposed survey of the whole state is merely an intelligent broadening of the planning which heretofore has been localized. . . . So far as I know this is the first time in the United States that the city or regional planning idea has been extended to take in a whole state. It will, therefore, be of great interest to everyone who realizes the importance of looking ahead and of using our resources to the best advantage" (Page 479, Vol. I, Papers and Addresses of Franklin Roosevelt).

In his annual message of January 7, 1931, Governor Roosevelt stated that a point had been reached where a definite far-reaching land policy should be formulated for the state; and on January 26, he sent a special message on land policy to the Legislature. The following quotations from this message indicate lines of thought which were repeated many times by the Governor in various public documents and speeches.

> What do we mean by this land policy? Fundamentally, we mean that every acre of rural land in the state should be used only for that purpose for which it is best fitted and out of which the greatest economic return can be derived.

> . . . it is unquestionably true that thousands of families, year after year, are spending labor and money in various parts of the state trying to get agricultural products out of land which will never be able to yield a profit in crops, but which should be devoted only to reforestation or recreational purposes.

> The program . . . can be used as the basis for planning future state and local developments which depend for their complete efficacy upon accurate knowledge of the proper setlling of population. For example, when we proceed to construct or improve roads through the rural areas of the state . . . we should know whether or not the land through which the roads pass will ultimately support the farm population . . . In the same way, our policy of establishing additional school facilities should be accurately guided . . . This conclusion is equally true in connection with electric power and telephone lines. (Page 480, Vol. I, Public Papers and Addresses of Franklin Roosevelt).

It is significant that Mr. Roosevelt's interest in planning had a strong rural flavor which came not only from his personal experience with farm operations in New York and Georgia. In a speech made in New York in December 1931, he refers to the fact that he had just returned from a holiday in Georgia, where he had been doing some planning "in the rural sense." He spoke informally about the days nearly twenty years ago when 'Mr. Charles Dyer Norton and my uncle, Mr. Frederic Delano, first talked to me about regional planning, for the City of Chicago.' He goes on to say "I think that from that very moment I have been interested not in the planning of any one mere city, but in planning in its larger aspects."

Mr. Roosevelt's travels to his "other home" in Georgia perhaps inspired the transplanting of the regiona planning idea from New York State to a southern region. In an address before a Conference on Regionalism at the University of Virginia in July 1931, he described the work that had been done in New York State and alluded to the fact that the same approach might help to solve the problems he had observed in Georgia. There were present at that Conference a numbe of southern leaders like Mr. Howard Odum, who had been developing the thought of regional development. There were also present many of the leaders in the planning movement, including such persons as Lewis Mumford, Stuart Chase, Louis Brownlow, Henry Wright, Clarence Stein, and others. There is no record of any direct connection between that Conference and the enmergence of the regional planning idea in the Tennessee Valley Authority Act, but certainly it provided fertile soil for the transplanting of this idea from New York, southward.

That this transplanting had taken place before President Roosevelt came to Washington is indicated by several of his writings. In his chapter on the Tennessee Valley Authority in the book, *On Our Way*, the President makes the following statements:

> It seemed wise at this time to commence a project which had no parallel in our history. It is true, that beginning ten or twenty years ago, movements had started in various parts of the country to encourage city planning . . . Gradually people began asking why should we not plan for the country districts as well as the city. As Governor of New York, I had sponsored a state-wide planning movement which had its foundation in a study of the problem of the use to which all land should be put. . .

> Before I came to Washington I had decided that for many reasons the Tennessee Valley – in other words, all of the watershed of the Tennessee River and its tributaries – would provide an ideal location for a land use experiment on a regional scale embracing many states. ...

> This plan fitted in well with the splendid fight which Senator Norris had been making for the development of tpower and the manufacture of fertilizer at the Wilson Dam properties . . .

> In enlarging the original objectives so as to make it cover the whole Tennessee Valley, Senator Norris and I undertook to include a multitude of human activities and physical developments.
> (Pages 53–54, President Roosevelt's book *On Our Way*.)

> In a speech at Montgomery, Alabama, on January 21, 1933, following inspection of the Muscle Shoals properties, Mr. Roosevelt said, "My friends, I am determined on two things as a result of what I have seen today. The first is to put Muscle Shoals to work. The second is to make Muscle Shoals a part of an even greater

development that will take in all of that magnificent Tennessee River from the mountains of Virginia down to the Ohio and the Gulf . . . Muscle Shoals gives us the opportunity to accomplish a great purpose for the people of many states, and indeed, for the whole Union. Because there we have an opportunity of setting an example of planning not just for ourselves but for the generations to come, tying in industry and agriculture and forestry and flood prevention, tying them all into a unified whole over a distance of a thousand miles so that we can afford better opportunities and better places for living for millions of yet unborn, in the days to come."

(Page 887, Vol I, Public Papers and Addresses of Franklin Roosevelt.)

Further comments on this same point are made in a note on TVA in Volume II of the President's Public Papers and Addresses (p. 123). He says there, "As Governor of New York I had sponsored and brought about a state-wide planning movement . . . Up to that time, although many cities had begun to plan their future growth, little on a very large scale had been done for the country areas. Before coming to Washington I had determined to initiate a land use experiment embracing many states in the watershed of the Tennessee River. It was regional planning on a scale never before attempted in history. I . . . planned for the development of the entire Tennessee Valley by a public authority similar to public authorities created in New York while I was Governor. . . . This plan . . . fitted in well with the project which had been urged for many years by Senator Norris . . . We planned to enlarge the project . . . to include a multitude of activities and physical developments."

The broadening of the Muscle Shoals development to include the whole river had occurred in earlier Norris bills. In 1928, Senator Norris stated (Cong. Record Vol. 69, p. 3441) . . . "It will be remembered that the several bills, modified in some respects, which I have been trying to have passed through the Senate, provided for the development of the Tennessee River and all its tributaries, from the mouth to the source of every one of the rivers." (Report #23, April 11, 1933, 73d Congress, 1st Session.) "The new provision of the bill" he states, "provides that for the proper use, conservation and development of the natural resources" of the Valley the President is authorized to make surveys and plans (as Outlined in Sec. 22). "In general the bill provides for a comprehensive method of development of the Tennessee Valley insofar as such development is of national concern."

The favorable report of the House Committee on Military Affairs made April 20, 1933, on H.R. 5081, which contained the same general provisions for regional investigation and development as S. 12723, quoted conservation sections of the different party platforms from 1912 to 1932 and commented as follows: (Report No. 48, 73rd Congress, 1st Session, April 20, 1933.)

The policy of this bill for the development of the Tennessee Valley is no new policy in American politics. For more than 20 years the conservation of our natural resources and their preservation for the use and benefit of all the people to whom they belong, both by the law of the land and the law of nature, has been a burning issue in the nation . . .

When this Tennessee Valley development shall have progressed sufficiently for us to learn great lessons as to how best to serve the public, then development will follow in other great inter-state and international water courses. Undoubtedly there are several great areas in all sections of the country that will ultimately be developed by the application of the same principles and policies.

The two ideas of developing the Muscle Shoals properties and inaugurating a broad program of regional planning for the Tennessee Valley had become united in the President's mind soon after his election if not some time before. With reference to this point, Senator Norris said, in discussing the TVA Bill on the floor of the Senate in May 1933, "The President made some suggestions before he was inaugurated in regard to enlarging somewhat the scope of the bill. I am not claiming any credit for myself, but it just happened that the enlargements suggested were things which I had never put into the bill but which I had openly, many times in the Senate and a great many times elsewhere, publicly advocated. I was glad that we were going to have a President who would make it possible to add something to the bill which would broaden its scope and, in my judgement, in the end, especially for future generations, be of very material value to the happiness and comfort of the people." (Page 2781, Congressional Record, Vol. 77.)

The merging of the President's thinking with that of Senator Norris is illustrated by a number of phrases in the TVA Act which hark back to the New York State experience. His interest in land utilization appears in the second phrase of the title, namely, an Act "To provide for reforestation and the proper use of marginal lands in the Tennessee Valley." This thought appears again in Section 23 where the President is directed to recommend to Congress legislation "for the especial purpose of bringing about in said Tennessee drainage basin and adjoining territory . . . (4) the proper use of marginal land; (5) the proper method of reforestation of all lands in said drainage basin suitable for reforestation." In the Norris Bill the above points (4) and (5) were followed by point (6), "the most practical method of improving agricultural conditions in the Valleys of said drainage basin." In the Act this phrase was changed to read, "the economic and social well-being of the people living in said river basin." The idea of retiring marginal lands from active use had been stressed by the President on many occasions, both for the purpose of improving the condition of people making a livelihood from the land and to reduce the cost of governmental services.

The message which President Roosevelt sent to Congress in April 1933 also illustrates the merging of his thoughts with those of Senator Norris. He first refers to the importance of enlisting the idle Muscle Shoals projects in the service of the people and then goes on with the familiar statement,

It is clear that the Muscle Shoals development is but a small part of the potential public usefulness of the entire Tennessee Rive. Such use, if envisioned in its entirety, transcends mere power development: it enters the wide fields of flood control, soil erosion, afforestation, elimination of agricultural use of marginal lands and distribution and diversification of industry. In short, this power development of war days leads logically to national planning for a complete river watershed involving many states and the future lives and welfare of millions. It touches and gives life to all forms of human concerns.

Although President Roosevelt had been a strong advocate of public power development while Governor of New York, and hence saw eye to eye with Senator Norris on the power features of the TVA Act, it is interesting that the President's first interest in the Tennessee Valley always appeared to be in the broader aspects of the project. In his message suggesting creation of the Authority he says first, that "It should be charged with the broadest duty of planning for the proper use, conservation and development of the natural resources of the Tennessee River drainage basis and its adjoining territory for the general social and economic welfare of the nation." He then says, "This Authority should also be clothed with the necessary power to carry these plans into effect."

There is no doubt that many leaders of thought in the subjects of land utilization and regional planning and development had a part in the evolution of the President's thinking, but it also seems clear from the record that the regional planning idea was basically the President's own, that it was he who introduced it to the Tennessee Valley legislation. It also seems clear that Senator Norris and President were in complete harmony as to their general objectives, and that therefore the merging of the two lines of thought was perfectly natural.

TBA:O

Source: *Personal files of A.J. Gray*

## MEMORANDUM

December 31, 1942
To: Mr. Howard K. Menhinick
From: Tracy B. Augur

ORIGINS OF SECTIONS 22 and 23 IN THE TVA ACT

The form and content of Sections 22 and 23 suggest that they had a different legislative history from that of other portions of the TVA Act. The remainder of the act deals with the execution of plans which had been formulated in greater or less degree throughout a long period of discussion about Muscle Shoals. Sections 22 and 23 deal with the making of plans as a basis for future legislation. The remainder of the act confers authority upon the TVA Board. Sections 22 and 23 confer authority on the President.

It is interesting also to compare the wording of the President's TVA message with the wording of Section 22. In his message the President proposed an Authority "charged with the broadest duty of planning for the proper use, conservation and development of the natural resources of the Tennessee River Drainage Basin and its adjoining territory for the general social and economic welfare of the nation." He added "This Authority should also be clothed with the necessary power to carry those plans into effect." In the Act itself, concern is centered on the execution of plans for river development, fertilizer manufacture, and power distribution, and the broad planning for the economic welfare of the region is added as something to follow and grow out of the other programs.

This difference is perhaps significant only as an illustration of the different approaches taken by the President and Senator Norris. Senator Norris had always been pressing for action to put the Muscle Shoals developments to use for the public benefit. The earlier legislation which he had introduced was concerned with fertilizer manufacture, the production and sale of low-cost power, and the full improvement of the river for power, flood control and navigation as an adjunct to the full use of the Muscle Shoals properties. Primarily, he appeared interested in getting something started in the realm of low-cost public power. Other things were added as matters of good political strategy or as outgrowths of the power programs. In this way Norris brought together the various lines of thinking that had previously grown up in relation to Tennessee River development, but his main interest in 1933 was in action. Plans for his action program had been pretty well prepared by the Army Engineers for the comprehensive river improvement.

The President was no less interested in action, particularly in the start of a public works program that would provide employment in 1933. However, his approach to the Tennessee Valley development was clearly influenced by his interest in broad planning for effective land use. In his book *On Our Way* he tells how he felt it wise "to commence a project which had no parallel in our history." He speaks of the growth of the planning movement from its original concern with cities, to metropolitan areas, and then to states. "As Governor of New York," he states, "I had sponsored a state-wide planning movement which had its foundation in the study of the problem of the use to which all land should be put. With this went, of course, the purpose of using land to the best advantage." His next passages are particularly significant. "Before I came to Washington I had decided that for many reasons the Tennessee Valley – in other words, all of the watershed of the Tennessee River and its tributaries – would provide an ideal location for a land use experiment on a regional scale embracing many States. In January I visited Muscle Shoals with a group of officials and experts, and subsequently announced plans for a comprehensive development of the entire Tennessee Valley region. These plans as developed contemplated the creation of a public authority to direct the development of a region comprising hundreds of thousands of square miles. This plan fitted in well with the splendid fight which Senator Norris had been making for the development of power and the manufacture of fertilizer at the Wilson Dam properties which had been erected by the Government during the World War. In enlarging the original objective so as to make it cover the whole Tennessee Valley, Senator Norris and I undertook to include a multitude of human activities and physical developments. By controlling every river and creek and rivulet in this vast watershed, and by planning for a highly civilized use of the land by the population of the whole area, we believed that we could make a lasting contribution to American life."

As Governor of New York, Mr. Roosevelt inherited the state planning activities begun under Al Smith. In many ways these had their origin in the conservation movement sponsored by Theodore Roosevelt and Gifford Pinchot, but they were also influenced by the city planning movement. The work was under the direction of Clarence Stein, and many others of the more progressive planners of that day were associated with it. In short, Roosevelt inherited what was undoubtedly the most comprehensive and best thought out planning activity developed up to that time.

Whether the New York State plan was "regional planning" in the modern sense is largely a question of terminology. New York State contained most of the rural and urban problems that any region is apt to have, and the scope of the New York State plans was certainly as broad as those for any region subsequently tackled. One of the principal concerns of the New York plan was to retire sub-marginal land from agricultural use, and Roosevelt pushed this program with a state bond issue to purchase such lands and convert them into public domain.

A reading of his speeches as Governor shows that the idea of using natural resources for public benefit was constantly in his mind and was growing in breadth. There is repeated reference to the proper utilization of land. It was only natural that a man with this breadth of interest in the development of a large territory should be inspired by the opportunity to inaugurate a similar program in a region such as the Tennessee Valley, and the social convergence of his views and those of Senator Norris might have come about from any of a dozen causes.

It seems certain that the Conference on Regionalism at the University of Virginia in 1931, had an important bearing. This conference was sponsored by the group of men who had been the progressive element in the planning movement, many of whom had been associated with the New York State plan. There were also present a group of southern leaders who had been developing the idea of regionalism. The conference was held in a southern environment and had a strong southern flavor. Mr. Roosevelt was the opening speaker and gave an excellent talk on what New York State had accomplished through state planning. In his talk he referred to the similarity between southern problems and those tackled in New York, and indicated that the same type of approach might be helpful. He mentioned specifically his "other home" in Georgia, the surplus production of peaches and pecans due to lack of proper planning, and the milk train that rumbled by each night carrying from Wisconsin to Florida milk which might better have been produced in the South.

Unfortunately, no notes were kept on the discussions which followed the Governor's speech, but he remained on the platform for nearly an hour answering questions and expanding on his views. I do not recall that the Tennessee Valley region was ever mentioned by name, but Mr. Odum and other southern regionalists were present and it is probable that the relationship of New York State planning techniques to the problems of the Southern Appalachians was mentioned. Following his conference talk, Governor Roosevelt had luncheon with leaders in the group and probably spent some time with them in the afternoon.

The Virginia Conference had a bearing on the TVA Act in another way. Actual authorship of Sections 22 and 23 is credited to John Nolen, Jr. and Frederick Gutheim, both of whom were present throughout the entire week of meetings at Charlottesville. They were both young men with active minds and sincerely devoted to the cause of regional planning. They were in contact with such men as Louis Brownlow, Lewis Mumford, Henry Wright, Benton MacKaye, Clarence Stein, Stuart Chase, Professor Odum, and other leaders of thought on the regional planning subject. Stuart Chase was at that time outlining a regional planning program for the Northwest along lines very similar to those in the Tennessee Valley.

I do not know to what extent the idea of Sections 22 and 23 originated with Messrs. Nolen and Gutheim and to what extent it was inspired by contacts with others at the Virginia Conference and in subsequent years, but there does seem to be a clear relationship between the President's appearance at Virginia and the final introduction of Sections 22 and 23. This relationship could doubtless be developed more clearly by correspondence with men like Brownlow, Gutheim, Odum and others who were present.

TBA:O
Tracy B. Augur

**MEMORANDUM**

March 1, 1943
To: Mr. Howard K. Menhinick
From: Tracy B. Augur

HISTORY OF SECTIONS 22 AND 23 OF THE TVA ACT

Correspondence with Mr. John Nolen, Jr., has brought out further points of interest about the origins of Sections 22 and 23. The first edition of the Tennessee Valley Act was submitted immediately after Congress convened in 1933 in the form of a Joint Resolution introduced in the Senate on March 9 by Senator Norris (S.J.Res.4). Sections 22 and 23 of this Resolution were the forerunners of Sections 22 and 23 in the TVA Act. Early in April Messrs. Gutheim and Nolen submitted an amplified version of these sections to Senator Norris and discussed it with him. This version greatly broadened the concept of Section 22 and introduced the phrases about preparing plans useful to Congress in the guidance of later development. A few days later, on April 11, Senator Norris submitted a Bill (S 1272) as a substitute for his previous Joint Resolution and containing substantially the revised wording, (as Sections 23 and 24 in this draft). When the Bill was reported back to the Senate from its Committee on Agriculture and Forestry later in April, Senator Norris suggested a few additions from the floor, which were accepted. These later additions to Section 22, permitting cooperation with local agencies, apparently were suggested by Dr. A.E. Morgan. It is not known whether he also suggested the revised wording of point (6) in Section 23. The wording in the Norris Bill was "The most practical method of improving agricultural conditions in the Valley of the said drainage basin."

For your information, the following material is attached.

1. Excerpts from letter of February 26, from John Nolen to me, marked personal, giving history of the revision.
2. Original wording of Sections 22 and 23 of Senate Joint Resolution 4, introduced by Senator Norris March 9, 1933.
3. Revised wording of Sections 22 and 23 as proposed by Messrs. Nolen and Gutheim, with additions to the Norris draft underlined.
4. Wording of Sections 22 and 23, as finally adopted in the TVA Act, with underlinings to show material suggested by Messrs Nolen and Gutheim (single underlining) and that added by Senator Norris on the floor, presumably suggested by Dr. A.E. Morgan (italicized).

Tracy B. Augur
TBA:REM
Encl.

EXCERPTS FROM LETTER DATED FEBRUARY 26, 1943
FROM JOHN NOLEN, JR. to TRACY B. AUGUR

Unfortunately my file of material on the T.V.A. bill did not include any diary of its development, so that I cannot give you full details. From the data in my files, however, it appears:

1. The original Norris bill was introduced as S.J. Res. 4 sometime in March 1933.
2. That Mr. Gutheim and I presented to Senator Norris on or shorly after April 5 the enclosed suggested amendment of Sections 22 and 23 of the original bill using, I believe, the accompanying notes in our oral presentation, the discussion lasting, as I recall, between 15 and 30 minutes.
3. That Senate 1272 was introduced as a substitute, by Senator Norris on April 11, and reported by the Committee on Agriculture and Forestry immediately thereafter with a 2-page Senate report No. 23, which merely describes the provisions of the new bill and makes no reference to the original draft.
4. That the amendments offered by Senator Norris on the floor were, so far as I know, suggested by Dr. Morgan of Antioch College.
5. That on April 11 Congressman McSwain introduced H.R. 4859, and Congressman Almon H.R. 4860, followed by H.R. 5081 introduced by Congressman Hill on April 20, which was the bill reported by the House on the same day with House report No. 48.
6. That finally Congressman McSwain submitted to the House report No. 130, being the conference report to accompany H.R. 5081, which was the bill finally passed, as you doubtless know.

No doubt the enclosed will answer the principal question that you raised as to the origin of Sections 22 and 23, but it may be that you wish to trace its further evolution in the subsequent bills, and particularly comment in the Committee and Conference Reports. If so, I would be very glad to lend you my copies of the above referred to bills and reports, and show you other materials the next time you are in Washington and we have an opportunity to go over them together.

Our part in the above was of course entirely unofficial, and ended with our conference with Senator Norris. Lest there might be some misunderstanding, I would appreciate it if you do not involve me in any official reports you make on the evolution of the bill.

TENNESSEE VALLEY DEVELOPMENT

Original Wording of Sections 22 and 23 of S.J. Resolution 4
Introduced by Senator Norris, March 9, 1933

Sec. 22. Within the limits of estimates made by the Bureau of the Budget and appropriations made therefore by Congress, the President is hereby authorized by such means or methods as he may deem proper, to make a survey of the said Tenessee Valley, including the valleys of its tributaries (1) for the purpose of reforestation of leands of said valleys; (2) for the purpose of proper use of marginal lands in said valleys; (3) for the location of dams upon said Tennessee River, or any of its tributaries, in order to control the flood waters and improve the navigability of the Tennessee River; (4) for the purpose of generating electricity at any or all of said dams; and (5) for the general purpose of improving the agricultural and other proper uses of the lands of said valleys.

Sec. 23. The President shall, from time to time, as the work provided for in section 22 progresses, recommend to Congress such legislation as he deems proper for the purposes of bringing about in said Tennessee Valley (1) the maximum amount of flood control; (2) the maximum development of said Tennessee River for navigation purposes; (3) the maximum generation of electric power consistent with flood control and navigation; (4) the proper use of marginal lands; (5) the proper method of reforestation of all lands in said valley suitable for reforestation; and (6) the most practical method of improving agricultural conditions in said valleys.

TENNESSEE VALLEY DEVELOPMENT

[(Proposed Revision of Sections 22 and 23 – S.J. RS. 4 (Sen. Norris)]

*(Changes from original proposal shown in underline)*

Sec. 22. To aid further the proper use, conservation and development of the natural resources of the Tennessee River drainage basin and of such adjoining territory in the same or nearby States as may be related to or materially affected by the developments consequent to this Act, and to provide for the general welfare of the citizens of said areas, the President is hereby authorized by such means or methods as he may deem proper within the limits of appropriations made therefore by Congress, to make such surveys and general plans for said Tennessee basin and adjoining territory as may be useful to the Congress and to the several States in guiding and controlling the extent, sequence and nature of development that may be equitably and economically advanced through the expenditure of public funds or through the guidance or control of public authority, all for the general purpose of fostering an orderly and proper physical, economic and social development of said areas; and the President is further authorized in making said surveys and plans to cooperate with the States affected thereby.

Sec. 23. The President shall, from time to time, as the work provided for in Section 22 progresses and indicates, recommend to Congress such legislation as he deems proper to carry out the general purposes stated in said section and for the especial purpose of bringing about in said Tennessee drainage basin in conformity with said general purposes (1) the maximum amount of flood control; (2) the maximum development of said Tennessee River for navigation purposes; (3) the maximum generation of electric power consistent with flood control and navigation; (4) the proper use of marginal lands; (5) the proper method of reforestation of all lands in said drainage basin suitable for reforestation; and (6) the most practical method of improving agricultural conditions in the valleys of said drainage basin.

## TENNESSEE VALLEY DEVELOPMENT

Wording Of Sections 22 And 23, as Finally Adopted in the TVA Act Of 1933

*(Changes from Sen. Norris proposal shown in Italics)*

Sec. 22. To aid further the proper use, conservation and development of the natral resources of the Tennessee River drainage basin and of such adjoining territory in the same or nearby States as may be related to or materially affected by the developments consequent to this Act, and to provide for the general welfare of the citizens of said areas, the President is hereby authorized by such means or methods as he may deem proper within the limits of appropriations made therefore by Congress, to make such surveys and general plans for said Tennessee basin and adjoining territory as may be useful to the Congress and to the several States in guiding and controlling the extant, sequence and nature of development that may be equitably and economically advanced through the expenditure of public funds or through the guidance or control of public authority, all for the general purpose of fostering an orderly and proper physical, economic and social development of said areas; and the President is further authorized in making said surveys and plans to cooperate with the States affected thereby, *or subdivisions or agencies of such States, or with cooperative or other organizations, and to make such studies, experiments, or demonstrations as may be necessary and suitable to that end*. (48 Stat. 69.)

Sec. 23. The President shall, from time to time, as the work provided for in Section 22 progresses and indicates, recommend to Congress such legislation as he deems proper to carry out the general purposes stated in said section and for the especial purpose of bringing about in said Tennessee drainage basin in conformity with said general purposes (1) the maximum amount of flood control; (2) the maximum development of said Tennessee River for navigation purposes; (3) the maximum generation of electric power consistent with flood control and navigation; (4) the proper use of marginal lands; (5) the proper method of reforestation of all lands in said drainage basin suitable for reforestation; and (6) *the economic and social well-being of the people living in said river basin*. (48 Stat. 69)

*Source: Personal files of A.J. Gray*

**MEMORANDUM**

October 3, 1933
To: Chairman A. E. Morgan, David E. Lilienthal
From: H.A. Morgan

PROPOSED STATEMENT OF POLICY IN THE PLANNING
ACTIVITIES OF THE TENNESSEE VALLEY AUTHORITY

The Tennessee Valley Authority has two kinds of functions: first, there are those relating to specific undertakings which are to be executed according to policies set out with some degree of definiteness in law. These are principally (a) the construction of the dam at Cove Creek, (b) the operation of the hydro-electric project, (c) the execution of the fertilizer program.

Under Section 22 of the Act, however, a different kind of function is involved. The President is authorized to make surveys and plans "as may be useful to the Congress and to the several States in guiding and controlling" the development of the area, and the President is authorized to make "such studies, experiments and demonstrations" as may be suitable. By executive order, the powers of the President have been delegated to the Tennessee Valley Authority under this section.

Section 22 raises sharply the question of the relation between the Authority on the one hand and people and their own agencies on the other.

There are two procedures which may be adopted in this program of studies and plans under Section 22. The Authority might carry out these studies exclusively through its own staff of experts, or, as an alternative, it could carry out these surveys and studies in collaboration with the existing agencies of the Valley area. If the latter is adopted, the problem of proper procedure would depend upon the kind of problem involved. In most cases the State and private universities including their research departments, State colleges of agriculture, their research experiment stations, and farmers' organizations such as farm bureaus and cooperative societies would collaborate with the Tennessee Valley Authority.

The proper procedure with respect to developing the interest of the people and their existing agencies in the Valley is as important in the process of making surveys and plans contemplated as in the execution of these plans. In other words, a policy with respect to the relationship between the people of the Valley and their agencies and the Authority may be essential as a basis for the studies and plans than if the Authority had been delegated by the President to make these studies independently as a basis for further Congressional or State Acts, under Section 22.

The Federal Government, for many years, has promoted the interest of the people of the States and the State agencies in a way similar to the suggested cooperation of the Authority, Federal departments have collaborated with the State authorities in developing and supervising projects that have been carried out to the people. This is well illustrated in the cooperation of Federal Department of Agriculture in agricultural research and extension. It is proposed that the Authority definitely adopt this method in carrying out the functions delegated to it by the President under Section 22.

If the surveys and plans are to have genuine importance, they must necessarily be such surveys and plans as can form the basis of action by the appropriate agency and can be actually carried into effect. It is suggested that the Board recognize in a statement of policy that these plans will be futile unless in their formulation the people of the Valley and their agencies have participated. For the Authority to proceed in the making of surveys and plans without such work being in collaboration with and in harmony with the existing agencies, will necessarily breed antagonism, distrust and a feeling that the years of work in this direction will be disregarded and cast aside.

It is implicit in Section 22 that the surveys and plans shall be the basis for legislative action. The Act says that the President is authorized "to make such surveys of and general plans for said Tennessee drainage Basin and adjoining territory <u>as may be useful to the Congress and to the several States</u> in guiding and controlling the extent, sequence, and nature of development that may be equitably and economically advanced through the expenditure of public funds or through guidance and control of public authority."

It is suggested further then, that the Authority definitely adapt a policy of preparing the surveys and plans in cooperation not only with the existing social, educational and economic agencies in the Valley, but also with the agencies of the State.

If this policy is adopted, then when the time comes for recommendations, such as the President is authorized to make under Section 23, to carry out the lessons learned from those surveys and to put the plans into effect, the people will feel that this phase of the project is theirs, they will be interested in it because they have been made a part of it and this being so, will aid in carrying the recommendations into action. If they are disregarded or their participation is only minor while the Authority dominates the picture, they will not feel that the project is theirs and will not have the same willingness to put the program through.

Specifically, it is suggested that the Authority adopt the following policy:

1. That the Authority will seek to stimulate and promote studies and surveys which have already been undertaken by the various agencies of the Valley.
2. That the Authority will stimulate and promote agencies of the Valley in the making of such surveys and studies as will permit the Authority to carry out its obligations under Section 22.
3. The Authority will not set up an organization to make studies and surveys which will oust or disregard existing agencies already in the field, willing and equipped to make such surveys and studies.

4.  The Authority will set up a sufficient staff to permit of the coordination and stimulation of existing agencies engaged in such surveys and studies.
5.  Where no existing Valley agency is available or can be set up, and only then, the Authority will set up an organization to develop studies, surveys and plans.

H.A. Morgan

**MEMORANDUM**

September 11, 1946
To: Paul W. Ager, Chief Budget Office
From: Tracy B. Augur, Assistant to the Director, Dept. of Regional Studies

DEVELOPMENT OF TENNESSEE VALLEY RESOURCES –
STATEMENT OF JULY 24, 1946

I have read the greater part of this statement and feel that on the whole it gives an excellent presentation of TVA programs and interests. I have only a few suggestions for changes and although I understand it is too late to incorporate these in the text being used this year, I am submitting them for whatever value they may have in future revisions.

The first relates to the text on Page 3. In this case you may wish to change the wording before presentation to the Budget Bureau this year because the statement appears to imply a much broader grant of developmental powers than is actually contained in the Act of claimed by TVA.

The paragraph in the middle of Page 3, as now worded, implies that Congress conferred on TVA authority to do the things required for the proper use, conservation, and development of the natural resources of the region and to provide for the general welfare of its citizens. The quoted parts of Sections 22 and 23 are lifted out of their context. Actually these sections do not confer authority to do anything about the development of resources except to make surveys and plans (and incidental experiments and demonstrations) and to recommend action. The powers granted by these sections of the Act are advisory powers and not developmental ones. The sections are planning sections and were so intended by their authors.

I think that the facts about the legal powers of TVA with respect to resource development are stated more accurately in the unpublished 1940 report "Regional Development in the Tennessee Valley". For your convenience the three opening paragraphs of that report are quoted herewith:

> The statute creating the Tennessee Valley Authority is a response to several interrelated problems traditionally national in character and interest. Far from being a novel excursion of the Congress of the United States, the Tennessee Valley Authority Act results from a legislative background of unusual maturity and is deeply rooted in a century of American history. The uniqueness of the statute is not in the problems toward which it was addressed, but in the method devised to meet these problems – that of a public corporation charged at once with the conduct of certain specific activities of the national government on a regional scale and with the responsibility of using this experience to assist in further regional development.
>
> The positive acts which the Authority is authorized to perform are specific, limited, and of long acceptance. It is directed to construct a series of dams on

the Tennessee River and its tributaries to make the 650 miles of the main river navigable and, at the same time, effective in flood control; to distribute and sell the water power created by these dams in amounts consistent with their operation for navigation and flood control; to utilize the Muscle Shoals properties for the experimental manufacture and distribution of fertilizers; and to maintain these properties in stand-by condition for national defense purposes. In addition to this specific program of action, the Authority is directed to make surveys, demonstrations, and recommendations for legislation in order to promote further the general development of the region and to cooperate towards this end with existing public agencies – local, state, and federal. In these latter aspects, it should be noted that the Authority is only an advisory and cooperating body. It can survey, demonstrate, and recommend to other interests properly authorized to act; it cannot, however, without congressional authorization execute its own recommendations.

In essence the Authority's specified operational functions constitute an integrated approach to the interrelated water-resource problems of an entire watershed. By and from a considered application of this integrated approach to water control in the Tennessee Valley, it has been possible to derive benefits even above those originally contemplated and to discover opportunities for further benefits which have been translated into the life of the region through cooperation with local, state, and federal agencies. In this manner, regional development beyond the utilization of the water reource is being fostered in the Tennessee Valley. It is a continuing process in which local institutions and state agencies cooperate with the federal government on problems of mutual concern. This unity of interest and effort leads, in a practical democratic fashion, to regional accomplishments not otherwise attainable.

Another statement about which I have some question is the one at the top of Page 2 that "The Tennessee Valley is predominantly an agricultural area." This is true if one is speaking in terms of predominant land use and if one construes that the operation of forests and woodlots is agriculture. It is also true if one is speaking in terms of the place of residence of the Valley population, although the 1940 census actually lists more dwelling units in the urban and rural non-farm category than in the farm category. (Apparently due to the large size of farm families, more than half of the Valley's people but less than half of the Valley's families live on farms.)

However, in economic terms the Tennessee Valley is not predominantly agricultural. Recent estimates by the U.S. Department of Commerce on the total returns from wages, salaries, and proprietor's income indicate that in the seven Valley states agriculture has been steadily losing ground over the past fifteen years and manufacturing has been gaining ground. In 1929 agriculture accounted for 23% of the total income but this had dropped to 20% by 1939 and 19% by 1944. On the other hand, manufacturing had risen from 18% of the total in 1929 to over 20% in 1939 and to 24% in 1944. If one discounts 1944 as a war year, the fact remains that in 1939 manufacturing had a higher rating, by a few percentage points, than agriculture.

In addition to the fact that manufacturing has achieved at least equal importance with agriculture in the Valley economy, it should be noted that the two together account for less than 50% of the Valley income. Trade and services account for a higher percentage of the total than either agriculture or industry.

The really important thing about the Tennessee Valley is that the long predominance of agriculture in the economy has ended and that the Valley is in the midst of a period of readjustment to a more diversified and better balanced economy. This does not mean that agriculture is not important but it does mean that it is no longer of predominant importance and that its relative importance is steadily declining.

I stress this point because I think that TVA is continually guilty of overstating the importance of agriculture and of basing its programs on past conditions rather than on those of the present and future. If TVA really wishes to assume a position of leadership in regional resource development, it should be giving much more attention to questions of industry, trade, and commercial services, which are the comng things in this region, and less attention relatively to agriculture.

This leads to another point which is not treated at all in your budget statement, namely, TVA's concern with the development of urban communities. It is in and around the urban communities that the new life of the region is developing. In a sense they are the basic resources which business and industry use and on their quality depends the economic future of the region. There is as much abuse and misdirected development of the urban resources of the Valley as there is of the agricultural resources and in contrast to agriculture, little or nothing is being done to stop it. The TVA program of community planning assistance is making a very meager entry into this exceedingly important field of resource development. There is not much that can be said about it but I think there should be at least some indication that TVA is aware of the problem.

CC to L.L. Durisch
A.J. Gray
J.P. Ferris

TBA:OM

# CHARTER OF THE REGIONAL RESOURCE STEWARDSHIP COUNCIL
## (*as of 2002*)

### Establishment, Purpose, and Scope

Under the TVA Act, the Tennessee Valley Authority has been charged with the wise use and conservation of the natural resources of the Tennessee River drainage basin and adjoining territory for the general purpose of fostering the orderly and proper physical, economic, and social development of the Tennessee Valley region. As the region has developed and the population grown, the stewardship of its natural resources has become both more complex and more important. TVA has always cooperated and worked closely with other public agencies and private entities that have responsibilities for and interest in the use and conservation of the region's natural resources. As competition for finite resources grows, fulfilling TVA's integrated resource stewardship mission will require increased cooperation with these other public agencies and private entities. It is in TVA's interest, and the interest of the public it serves, to establish a mechanism for routinely obtaining the views and advice of the public agencies and private entities involved in and affecting natural resources stewardship. Accordingly, TVA establishes the Regional Resource Stewardship Council to provide TVA advice on its stewardship activities and the priorities among competing objectives and values. TVA's stewardship activities include the operation of its dams and reservoirs, its responsibilities for navigation and flood control, and the management of the lands in its custody, water quality, wildlife, and recreation.

### TVA Legal Responsibilities

In accordance with the Tennessee Valley Authority Act of 1933 and the Federal Advisory Committee Act, the purpose of the Council is to provide advice only, and TVA retains sole responsibility for the management and operation of its stewardship activities and for all decisions regarding matters under consideration by the Council.

### Term of Council

In accordance with the Federal Advisory Committee Act, the Council will terminate two years from the date this charter is filed with Congress unless it is renewed.

## Membership

The Council shall consist of up to 20 members appointed by TVA. All members of the Council shall be persons possessing demonstrated professional or personal qualifications relevant to achieving TVA's stewardship mission. The TVA Board shall ensure that the membership of the Council is balanced and that it represents and includes a broad range of diverse views and interests, including recreational, environmental, industrial, business, consumer, educational, and community leadership interests.

The Governors of Alabama, Georgia, Kentucky, Mississippi, North Carolina, Tennessee, and Virginia will each be asked to nominate a member to the Council, taking into account the need for a balanced and diverse membership.

TVA shall appoint seven members nominated by the Governors and shall appoint up to 13 additional members. The membership of the Council shall include at least four members representing distributors of TVA power and at least one member representing each of the following interests: a directly served customer of TVA, a beneficiary of TVA's navigation program, a beneficiary of TVA's flood control program, a recreational interest, and an environmental interest. TVA will appoint up to four additional members to ensure a balanced representation of a broad range of views.

Each member shall serve without compensation and shall not be considered an employee of TVA. Each member shall be appointed for a term of two years beginning with the charter date. Whenever a vacancy occurs, TVA may appoint a replacement for the remainder of the applicable term. TVA shall designate one Committee member as Committee Chair.

## Designated Federal Officer

The River System Operations & Environment Executive Vice President shall serve as the Designated Federal Officer for the Council (DFO). The DFO (or a designated substitute) shall ensure that proper public notice is given of each meeting of the Council, approve the proposed agenda for each meeting, attend each meeting, ensure that detailed minutes are taken at the meeting, ensure that the minutes and other Council records are available to Council members and the public, ensure that adequate facilities are provided for Council meetings and other needs, and make such reports about the operation of the Council as may be required or desirable.

## Meetings

The Council shall meet at least twice annually at the call of its Chair, with the concurrence of the DFO. Eleven voting members shall constitute a quorum for the conduct of business, and any recommendation by the Council to TVA shall require an affirmative vote of at least a majority of the total Council membership on that date. Each Council member shall be provided the opportunity to include minority or dissenting views to accompany recommendations by the Council to TVA. Notice of each meeting shall be provided in the Federal Register at least 15 days before each meeting. Each meeting shall be open to the public, unless closed in accordance with the Federal Advisory Committee Act and the Government in the Sunshine Act. The DFO and the Chair shall agree on the proposed agenda for each meeting sufficiently in advance of the meeting so that the agenda can be included in the Federal Register notice. Interested members of the public may attend meetings and file statements with the Council and, if permitted by the Chair and in accordance with Council procedures, may speak at a meeting. Meetings may be adjourned by the Chair upon approval by a majority or by the DFO.

## Subjects Council Shall Address

The DFO, in coordination with the Council Chair, will submit current issues for the Council's consideration. The Council may choose to address these topics or recommend consideration of other issues.

## Funding And Support Services

TVA shall provide the Council with sufficient facilities in which to conduct its meetings and to provide a repository for its minutes and other records. TVA will also provide the Council with appropriate clerical support as needed. TVA shall provide such additional funding as reasonably necessary to achieve the purposes for which the Council was created and shall provide any further guidelines and management controls as may be necessary to further the objectives of the Council. The estimated annual operating costs of the Council are $120,000 and TVA staff time equal to 1.0 full-time equivalent.

## Travel Reimbursement

Each member shall be entitled to reimbursement for travel and subsistence expenses which are incurred in connection with the attendance of Council meetings or otherwise incurred while engaged in the performance of Council duties and approved by TVA. Such reimbursements shall be subject to and computed and paid in accordance with Federal travel regulations and TVA procedures.

**Conflicts of Interest**

Members shall be considered representatives of the group, organization, or other entity identified by TVA in making the appointment. Members shall not, except in connection with such representation (or as otherwise disclosed to the Council) either (a) participate in Council matters affecting a personal interest of the member or the interest of a person or organization with which the member is closely affiliated, or (b) accept any gift from any person or organization having an interest in matters coming before the Council. Members shall not use Council resources for unauthorized purposes. In addition, the Ethics in Government Act of 1978 and any other conflict of interest statutes shall apply according to their provisions.

**Subcommittees**

TVA will create any subcommittees which may be necessary to fulfill the Council's mission.

# Bibliography

Allbaugh, L.G. (1953). *How TVA Fertilizers Are Used*, Florida State Lectures. TVA Technical Library, Knoxville.

Aguar, C.E. (1995) 'Earle S. Draper: His TVA Staff and Neighbors in Norris, Tennessee', in Proceedings, Sixth National Conference on American Planning History, Society for American City and Regional Planning History, pp 45–61.

Augur, T. (1936). The New Town of Norris, *The American Architect*.

Augur, T. (1940). *Regional Development in the Tennessee Valley*. Tennessee Valley Authority, Knoxville.

Birkhead, G.S. (1962). *Government in Metropolitan Areas: Commentaries on a Report by the Advisory Commission on Intergovernmental Relations, 87th Congress, First Session, December 1961*. US Government Printing Office, Washington, DC, pp. 87–93.

Bradshaw, M. (1988). *Regions and Regionalism in the United States*. University Press of Mississippi, Jackson and London.

Cartwright, J.M. (1979). *Role of the Planner in Comprehensive Community Energy Management*. Tennessee Valley Authority Division of Navigation Development and Regional Studies, Knoxville, Tenn., 29 pp.

Chandler, W.U. (1984). *The Myth of TVA : Conservation and Development in the Tennessee Valley, 1933–1983*. Ballinger Pub. Co., Cambridge, Mass., xvi, 240 pp.

Clapp, G.R. (1955). *The TVA: An Approach to the Development of a Region*. University of Chicago Press, Chicago.

Creese, W.L. (1990). *TVA's Public Planning: The Vision, the Reality*. The University of Tennessee Press, Knoxville, 388 pp.

Cutler, P. (1985). *The Public Landscape of the New Deal*. Yale University Press, New Haven and London, 182 pp.

Dahir, J. (1950). *Communities for Better Living; Citizen Achievement in Organization, Design and Development*, Harper, New York, 321 pp.

Davidson, D. (1948). *The Tennessee: Volume II, The New River; Civil War to TVA*. Rinehart & Company, Inc., New York, 177 pp.

Derthick, M. (1974). *Between State and Nation: Regional Organizations of the United States*. The Brookings Institution, Washington, DC, 242 pp.

Doig, J.W. (2001). *Empire on the Hudson: Entrepreneurial Vision and Political Power at the Port of New York Authority*. Columbia University Press, New York, 582 pp.

Draper, E. (1933). *unpublished news release*. Tennessee Valley Authority, Knoxville.

Draper, E. (1983). *Address on the occasion of the City of Norris Fiftieth Anniversary Celebration, October 14, 1983*.

Draper, E. (1984). *Interview*, TVA Oral History Collection, Knoxville.

Droitsch, D. (1994). *TVA's Blighted Nuclear Romance*, pp. 906–08

Durisch, L.L. (1951). *Local Flood Damage Control Problems and the Work of Local Planning Agencies*. In: W.J. Hayes, (ed.), Knoxville.

Durisch, L.L. and Hershal, L. (1951). *Upon Its Own Resources*. Tennessee Valley Authority, Knoxville.

Durisch, L.L. and Lowry R.E. (1953). 'The Scope and Content of Administrative Decision – The TVA Illustration'. *Public Administration Review*, pp.220 ff.

Fenneman, N. M. (1946) *Physical Divisions of the United States: Prepared by Neven M. Fenneman in cooperation with the Physiographic Committee of the Geological Survey*, Washington, DC: US Geological Survey.

Friedmann, J.R. (1955). *The Spatial Structure of Economic Development in the Tennessee Valley, a Study in Regional Planning*. University of Chicago, Department of Geography, Chicago.

Friedmann, J.R. and Weaver, C. (1979) *Territory and Function: The Evolution of Regional Planning*, University of California Press, Berkeley and Los Angeles.

Galbraith, J.K. (1994). *A Journey Through Time*. Houghton Mifflin Co., Boston and New York.

Grant, N. (1990). *TVA and Black Americans: Planning for the Status Quo*. Temple University Press, Philadelphia, xxxi, 207 pp.

Gray, A.J. (1956). 'Planning for Local Flood Damage Prevention'. *Journal of the American Institute of Planners*, pp.11–16.

Gray, A.J. (1957). *Urbanization: A Fact – A Challenge*, The Tennessee Planner, pp. 145–55.

Gray, A.J. (1958). *Upper French Broad Flood Control Project*. In: Durisch, L.L.(ed.), Knoxville.

Gray, A.J. (1961). "Communities and Floods". *National Civic Review*, pp.134–38.

Gray, A.J. (1962). *An Approach to Urban-Industrial Problems in the Tennessee Valley*. In: Durisch, L.L.(ed.), Knoxville.

Gray, A.J. (1966). *North Carolina – Development of the Upper French Broad*. In: Stern, P.M., Director of Regional Studies, TVA(ed.), Knoxville.

Gray, A.J. (1974). "The Maturing of a Planned New Town: Norris, Tennessee". *The Tennessee Planner* 32, pp.1–25.

Gray, A.J. (1988). *Interview with John Bonum*.

Gray, A.J. and Johnson, D.A. (1987). *TVA's Formative Years: The Planners' Contributions*, Second National Conference on American Planning History, Columbus, Ohio.

Gray, A.J. and Roterus, V. (1960). *The Tennessee River Valley: A Case Study*. Office of International Housing, Office of the Administrator, US Housing and Home Finance Agency, Washington, DC.

Hargrove, E.C. (1983). 'The Task of Leadership: The Board Chairmen'. In: Hargrove, E.C., Conkin, P.K. (eds.), *TVA Fifty Years of Grass-Roots Bureaucracy*. University of Illinois Press, Urbana.

Hargrove, E.C. (1994). *Prisoners of myth : the leadership of the Tennessee Valley Authority*, 1933–1990. Princeton University Press, Princeton, N.J., xvi, 374 pp.

Hargrove, E.C. and Conkin, P.K., (1983). *TVA Fifty Years of Grass-Roots Bureaucracy*. University of Illinois Press, Urbana, xvii, 345 pp.

Hirsch, S.P. (1979). *Status Report on Office of Community Development*. In: Freeman, S.D.(ed.), Knoxville.

Jacobs, J. (1984). *Why TVA Failed*, The New York Review of Books, pp. 41–47.

Jensen, M. (1965). *Regionalism in America*. The University of Wisconsin Press, Madison and Milwaukee, 425 pp.

Johnson, D.A. (1996). *Planning the Great Metropolis: The 1929 Regional Plan of New York and Its Environs*. E. & F. Spon, London.

King, J. (1959). *The conservation fight, from Theodore Roosevelt to the Tennessee Valley Authority*. Public Affairs Press, Washington, 316 pp.

Leiserson, A. (1983). 'Administrative management and political accountability'. In: Hargrove, E.C., Conkin, P.K. (eds.), *TVA Fifty Years of Grass-Roots Bureaucracy*. University of Illinois Press, Urbana.

Lepawsky, A. (1949). *State Planning and Economic Development in the South*. Kingsport Press, Kingsport, Tennessee.

Levin, M.R. (1968). 'The Big Regions'. *Journal of the American Institute of Planners*.

Lilienthal, D.E. (1945). *TVA : Democracy on the March*. Pocket Books, New York, vi, 247 pp.

Lorenz, J.D. (1960). *Arthur Morgan and the Tennessee Valley Authority*. Harvard University, Cambridge, Massachusetts.

MacKaye, B. (1935). *Physical Planning in the TVA Program*. In: Draper, E. (ed.), Knoxville.

Martin, R. (1956). *TVA, The First 20 Years, A Staff Report*. University of Alabama Press and University of Tennessee Press.

McCraw, T.K. (1971). *TVA and the power fight, 1933–1939*. Lippincott, Philadelphia, xi, 201 pp.

McCraw, T.K. (1978). *Morgan vs. Lilienthal: The Feud Within the TVA*. Loyola University Press, Chicago.

McDonald, M.J. and Muldowny, J. (1982). *TVA and the dispossessed: the resettlement of population in the Norris Dam area*. University of Tennessee Press, Knoxville, xv, 334 pp.

Menhinick, H.K. (1946). "Supreme Court Decision Assures Addition to Great Smoky Mountains National Park". *Planning and Civic Comment*, pp. 46–48.

Miller, H.V. (1987). *I Was There!*, unpublished memoir, 122 pp.

Moffett, M. and Wodehouse, L., (1983). *Built for the people of the United States : fifty years of TVA architecture*. Art and Architecture Gallery, University of Tennessee, Knoxville, 65 pp.

Moore, J.R., (1967). *The Economic impact of TVA*. University of Tennessee Press, Knoxville, xv, 163 pp.

Morgan, A.E. (1974). *The making of the TVA*. Prometheus Books, Buffalo, N.Y., xiv, 205 pp.

Morgan, E. (1991). *Arthur Morgan Remembered*. Community Service, Inc., Yellow Springs, Ohio.

Morgan, H.A. (1933). *Proposed Statement of Policy in Planning Activities of the Tennessee Valley Authority*. In: Morgan, A.E. (ed.), Knoxville.

Munzer, M.E. (1969). *Valley of Vision: The TVA Years*. Alfred E. Knopf, New York, 199 pp.

National Resources Committee (1935). *Regional Factors in National Planning*. US Government Printing Office, Washington, DC, 223 pp.

Nolt, J., Bradley, A.L., Knapp, M. Lampard, D.E., and Scherch, J. (1997). *What Have We Done?* Earth Knows Pulblications, Washburn, Tennessee, 303 pp.

Pritchett, C.H., (1943) *The Tennessee Valley Authority: A Study in Public Administration*, The University of North Carolina Press, Chapel Hill, 333 pp.

Pulsipher, A. (1984). *TVA's New Critics – Did They See the Play?*, Inside TVA.

Roberts, E. (1955). "One River – Seven States: TVA-State Relations in the Development of the Tennessee River". *The University of Tennessee Record* XXXI, pp.100.

Roosevelt, F.D. (1931). *Address to Roundtable on Regionalism*, University of Virginia, Charlottesville, Virgina.

Roosevelt, F.D. (1936). *Papers and addresses of Franklin D. Roosevelt: Volume I: The Genesis of the New Deal*. Random House, New York.

Schaffer, D. (1984). *Environment and TVA : toward a regional plan for the Tennessee Valley, 1930s*. Tennessee Valley Authority Cultural Resources Program, Norris, Tenn., 34 pp.

Selznick, P. (1949). *TVA and the Grass Roots; a Study in the Sociology of Formal Organization*. University of California Press, Berkeley, viii, 274 pp.

Selznick, P. (1966). *TVA and the Grass Roots; as Study in the Sociology of Formal Organization*. Harper & Row, New York, xvi, 274 pp.

Smith, F.E. (1966). *The Politics of Conservation*. Harper & Row, New York, 338 pp.

Talbert, R., Jr. (1987). *FDR's Utopian, Arthur Morgan of the TVA*. University of Mississippi Press, Jackson and London, 218 pp.

Tellico Reservoir Development Agency (1992). *Decade of Progress*, 6pp.

Tennessee State Planning Commission (1960). *Melton Hill Reservoir, Comprehensive Plan for Land Use Development*. Tennessee State Planning Commisssion, Nashville.

Tennessee Valley Authority. (1936). *Report to the Congress on the Unified Development of the Tennessee River System.* Tennessee Valley Authority, Knoxville.

Tennessee Valley Authority. (1938). *The scenic resources of the Tennessee Valley. A descriptive and pictorial inventory.* U.S. Govt. Print. Off., Washington, DC, xii, 222 incl. front., illus., 7 fold. maps, diags. pp.

Tennessee Valley Authority. (1949). *Watts Bar Steam Plant Studies.* Tennessee Valley Authority, Knoxville.

Tennessee Valley Authority. (1951). *Annual Report for the Fiscal Year Ending June 30, 1951.* Tennessee Valley Authority, Knoxville.

Tennessee Valley Authority. (1952). *Annual Report for the Fiscal Year Ending June 30, 1952.* Tennessee Valley Authority, Knoxville.

Tennessee Valley Authority. (1953). *Annual Report for the Fiscal Year Ending June 30, 1953.* Tennessee Valley Authority, Knoxville.

Tennessee Valley Authority. (1954). *Annual Report for the Fiscal Year Ending June 30, 1954.* Tennessee Valley Authority, Knoxville.

Tennessee Valley Authority. (1961). *Annual Report for the Fiscal Year Ending June 30, 1961.* Tennessee Valley Authority, Knoxville.

Tennessee Valley Authority. (1972). *TVA Handbook.* Tennessee Valley Authority, Knoxville.

Tennessee Valley Authority. (1972). *Annual Report.*

Tennessee Valley Authority. (1974). *Environmental Statement,* Elkmont Rural Village.

Tennessee Valley Authority. (1976). *Timberlake New Community General Land Use Plan.* Tennessee Valley Authority, Knoxville, pp. 33.

Tennessee Valley Authority. (1980). *Annual Report.* Tennessee Valley Authority, Knoxville.

Tennessee Valley Authority (1980) *Strategies for the 1980's: A TVA Statement of Corporate Purpose and Direction,* Knoxville.

Tennessee Valley Authority, (1982), *TVA: an Agency for Regional Development, 1982,* Knoxville.

Tennessee Valley Authority, (1986) *Multi-year Plan for 1987–1992, Tennessee Valley Authority,* Knoxville.

Tennessee Valley Authority (1987) *Options for Structure and Organization of TVA,* Knoxville. 83pp.

Tennessee Valley Authority (1988) *Managing Energy and Resource Efficient Cities from Concept to Impacts, A Report to the United States Agency for International Development,* TVA MEREC Staff, Knoxville, TN, April 1988, 48 pp.

Tennessee Valley Authoity (1989) *Integrated Regional Resources Management – Based on the Experience of the Tennessee Valley Authority,* Knoxville, 501 pp.

Tennessee Valley Authority (1989) *Managing Energy and Resource Efficient Cities (MEREC) Summary and Evaluation, 1981–1989: A Report to the*

*United States Agency for International Development*, TVA MEREC Staff, Knoxville, TN, September 1989, 31 pp.

Tennessee Valley Authority. (1994). *Final Environmental Assessment, Morgan Square Project, Greene County*. Tennessee Valley Authority, Knoxville.

Tennessee Valley Authority, *Inside TVA*, (various issues).

Thurman, S. (1986). *A History of the Tennessee Valley Authority*. Tennessee Valley Authority Information Office, Knoxville, 48 pp.

US Bureau of the Census (1930, 1940, 1950). *Census of Population*. US Government Printing Office, Washington, DC.

Van Fleet, A.A. (1987). *The Tennessee Valley Authority*. Chelsea House Publishers, New York, 92 pp.

Von Eckardt, W. (1965). *The Case for Building 350 New Towns*, Harper's Magazine, pp. 85–96.

Wagner, A.J. (1965). "Natural Resources: A Challenge for Planning". *The Tennessee Planner*, pp.71–72.

Wengert, N.I. (1952). *Valley of tomorrow; the TVA and agriculture*. Bureau of Public Administration, University of Tennessee, Knoxville, xv, 151 pp.

Wheeler, W.B. and McDonald, M.J. (1986). *TVA and the Tellico Dam, 1936–1979 : a bureaucratic crisis in post-industrial America*. University of Tennessee Press, Knoxville, xii, 290 pp.

Wildavsky, Aaron (1976). *Dixon-Yates: A Study in Power Politics*. Greenwood Press, Westport, CT.

# Index

Note: bold entries refer to illustrations.